Auto-Architect

The Autobiography of Gerald Palmer

(1911-1999)

with additional material
by Christopher Balfour

Magna Press

For Diana

Published by **Magna Press**
28 Allen Road, Great Bookham,
Nr. LEATHERHEAD,
Surrey KT23 4SL.

Direct sales enquiries 01372 451124

ISBN 0 9543121 1 2
First published 1998
Second revised and enlarged edition 2004

Copyright © 2004 Estate late Gerald Palmer,
Christopher Balfour, and Magna Press.

All rights reserved. No part of this publication may be reproduced, stored in a retrieval system, or transmitted, in any form or by any means, electronic, mechanical, photocopying, recording or otherwise, without the prior written permission of Magna Press.
In this book manufacturers' model names and logos have been used for purposes of identification and illustration. Such names are the property of the trademark holder and the MG name and octagon are acknowledged as the property of MG Rover.
All the photographs come from the author's collection, family albums, the Magna Press archives, and from Noel Stokoe as Jowett Historian. It has not been possible in every case to trace the original photographer and if any copyright has been infringed it has been done unintentionally and the sincerest of apologies are offered.

Printed and bound by Lightning Source

Contents

African Childhood	7
Scammell and O.D. North	13
The Deroy Adventure	25
Nuffield and the Oxford Vaporiser	31
Designing the Javelin	37
Producing the Javelin	49
Monte Carlo and the Return to Oxford	61
Nuffield again: Magnettes and Others	67
Pathfinder and Six/Ninety for B.M.C.	81
Wansbrough's Yeoman and Orchard House	93
Chief Engineer and the Clash with Leonard Lord	99
General Motors Experiences	109
Victor and Viva	117
Bugatti and the Work with F. J. Payne	125
Targa Florio Mercedes	129
Epilogue	137
Index	147

Acknowledgements

It is an honour to have worked with Gerald on this project. There is also a lesson about human life highlighted by Gerald's generosity of spirit. He has never complained about his treatment by Leonard Lord. (A passing sadness about the fate of the industry, but no personal resentment). He has always taken the view that you cannot ordain your working colleagues. If it had not been Lord, he could have clashed with someone else. Thus he did not, and does not, look back; grateful and glad that he was able to accept and enjoy other opportunities.

I am indebted to those in the Industry who have shared their memories, and to those who have recounted their experiences with Gerald's cars. I was able to relay the knowledge gained which acted as a catalyst to Gerald's impressive memory.

We thank you all, also those who have helped in other ways.

Ann Balfour - typing, E.Chalenor Barson - Deroy, Paul Batho - M.G., Stuart Broatch - Vauxhall, John Burman - Wolseley in Australia, Pam Chase - Javelin, Giles Chapman, Elaine Cowley - Vauxhall, Michael Edwardes - B.M.C., Gavin Farmer, Val Foster - Wolseley, Alan & Gill Goodyear - Vauxhall & Bugatti lamps, Michael Graves-Morris - Vauxhall, Mike Hawke - M.G., Brian Heath, Fred Holmes - Wolseley in Australia, Arthur Jeddere Fisher - Mercedes, Ann Joy - Javelin, Warren Marsh - M.G.Magnette, Geoff McAuley - Javelin, Kate Morley, Hubert Patthey - Javelin, Celia Palmer, Jon Pressnell, John Reilly - Javelin, Peter & Joy Richardson - Wolseley in Australia, Edward Riddle - North Lucas, Bernard Ridgeley - Vauxhall, David Rowlands - Riley, Dennis Sherer - Vauxhall, Noel Stokoe - Javelin, Peter Tothill - M.G., Riley, Wolseley, Peter & Val Upton, David Wansbrough - Javelin, Yeoman, Mike Worthington-Williams.

Christopher Balfour.
Winchester 1998

Preface to First Edition

Having spent the whole of my working life in the motor industry, been involved with several of its products, and worked with many of its talented leaders, friends have suggested that I record a few notes on my experiences. After some persuasion I agreed, and the result is this volume of reminiscences as I and others working in this field saw it.

My interest and enthusiasm has been entirely in the field of product design. So keen was I to pursue it that, as a teenage youth I came from Central Africa to an apprenticeship in England and retired after about fifty years in the industry. In a chequered career I held various positions, the highest as Technical Director of one of Britain's largest companies.

It was a period of great activity and innovation, of hopes unfulfilled, of early companies going out of business or amalgamating with others, and of new companies being formed. After an initial upsurge, the domestic industry declined, partly - perhaps it must be admitted - from self inflicted wounds. Important amongst these was the lack of investment in an industry which had become increasingly capital intensive as a result of the essential adoption of the all-steel body. There was also insufficient rationalisation of models, insufficient attention to test and development, and excessive demands from labour. These wounds and the intense competition from other countries inevitably led to increased penetration by American, European, and Far Eastern interests and personnel into most branches of the industry.

I am deeply indebted to Christopher Balfour in his capacity as advisor, for the interest, enthusiasm, hard work and guidance which he has unstintingly given me and which have brought this book to fruition. Without him I could not have done it. The same applies to Elizabeth Leigh, my neighbour, who did all the initial typing and secretarial work.

I hope you, reader, will find in it something of interest.

Gerald Palmer
Iffley, Oxford, 1998

Preface to Second Edition

There has been international response to Auto-Architect, from owners and admirers of the Javelin and Magnette in Britain, Europe and America, and also from Australia where they cherish the larger Wolseleys and Rileys which have survived in that drier climate. The Javelin story is well covered in other books. For Nuffield and BMC the Issigonis bandwagon rolls on, and it seems important to keep references to other sides of the story in print for future historians of the industry. Already, as seen in requests to the Society of Automobile Historians for information, students from other countries are researching for their projects the reasons for the decline of the British motor industry.

Many of Gerald's colleagues are no longer with us, but, in writing the epilogue, I have been greatly helped by correspondence with Jim O'Neill, Gerald's senior body engineer. He retains fond memories of his former boss and has helped fill out more details of the relationship with Issigonis. Jon Pressnell has again shared memories of his own talks with Gerald. There have been further conversations with David Rowlands who is still, after so many years, enjoying his two Pathfinders. With his help, and reference to Peter Thornley's book about his father, Mr. MG, I hope that car's difficult gestation is explained in a way which will satisfy the owners of the 300 odd examples that remain on Gordon Webster's RMH database. Jack Daniels, still going strong in his nineties, has confirmed the events recalled by Gerald of his last year at BMC. He wrote also of the proposal for a twin camshaft cylinder head for the C-series engine saying that this would have been both very expensive and built in small numbers and was 'for this reason cancelled'. He also said that Gerald had nothing to do with the FWD Minor project, but that the prototype had been driven by Lord and Harriman and that this experience may have influenced their decision to later accept the Mini concept.

The first hardbound, limited edition version of Gerald's autobiography is long out of print and the opportunity has been taken to reprint this in a less expensive format with his original text unaltered, but with some extra photographs. However, Gerald died in 1999 and I have taken the liberty of adding an epilogue where I am able to reflect on his work and perhaps add a little extra detail to the words he wrote towards the end of his life.

Christopher Balfour
Winchester 2004

Chapter One

African Childhood

'YOU ARE TWO CLEVER LITTLE BOYS – I USUALLY HAVE DIFFICULTY IN STARTING IT – HERE IS SIXPENCE FOR EACH OF YOU.'

'It' was a 1912 Austin 20 belonging to the assistant engineer in my father's office, outside of which he parked it all day. The little boys were me and my older brother and we had the naughty habit of clambering all over the car, sitting in the driver's seat pretending to drive it, as small boys do. On this particular morning we pressed and pulled various knobs and, much to our fright, the engine started. Hearing it, the owner came rushing out of the office and seeing no damage was done, took a very understanding attitude, scolded us mildly and told us not to do it again. That was my first encounter with the workings of a motor car.

The office was that of the district engineer of the Beira, Mashonaland and Rhodesia Railway in Umtali, Southern Rhodesia. My father was that district engineer and was responsible for the track from Salisbury, the capital, to Beira on the coast of Mozambique (Portuguese East Africa). During the First World War this railway was the vital link ferrying English and South African soldiers and equipment to the East African war and the costly campaign against General Van Lettow's guerillas in Tanganyika, then a German protectorate. They could not land further up the coast at Dar-es-Salaam as this port was securely held by the Germans. There was torrential rain in 1916 and one night the swollen river Odzi (inland from Umtali) swept away the steel railway bridge. There was overbearing pressure on my father to achieve a speedy repair and no suitable materials were readily to hand. (A wooden replacement would not have had sufficient strength.) The solution was to dismantle the structure of another bridge on a less important branch line near Beira.

I was then six years old and the picture is still clear in my mind of these two enormous girders being hauled up through Umtali on flat platform trucks. Father also took my brother and I up to the reconstruction site where pontoons were being used to carry the most urgent supplies across the swirling, reddish-brown torrent. The river bank had crumbled where the old bridge had been supported and I watched as the Mashona workers poured the concrete into the

Auto-Architect

steel caissons which would provide new supports. After months of toil the Beira girders were pushed out over the gap and the new track laid. When the time came for the trial crossing the great steam locomotive so dwarfed the minimal new structure (there was no side rail or excess ironwork) that it looked to us on the bank that the machine was flying unsupported above the waters.

Our dwelling house was next door to the office, about 100 yards away. Both were of wood and corrugated iron construction, raised on iron stilts to protect them from snakes, ants, and other nasty insects. However, neither the stilts, nor the voluminous nets which were draped around our beds, saved us from the malarial bites of the local mosquitoes. Too often we were confined to bed whilst the fever sweated out of our systems. The rail tracks of Umtali station were in front of these two buildings with a road in between. Some nights, when I couldn't sleep, I'd lie watching the beam of a searchlight fitted to the front of the locomotives dart across the wall of the room. These powerful lights could pick out the form of animals wandering onto the line way ahead of the train. On the other side of the tracks was the yard, the repository of all the timber and other materials used in keeping the track in good repair. Additionally, there were inspection trolleys of three kinds – push trolleys, pump trolleys, and a solitary motor trolley.

Above: Formal portrait of the Palmer family. Will, my father, is seated on the left with me standing between him and my sisters, Marjorie and Joan. My Mother, Esther, and elder brother, Ron, are on the right.
Opposite: My Father with the the five-man crew that provided the motive power to propel this Rhodesia Railways pump trolley.

Auto-Architect

The first were simple wooden platforms mounted on two axles, pushed along by two or more natives who, on a down grade, jumped on and had a ride. The second were flat wooden platforms mounted on two axles but with a superstructure on which were mounted two rocking arms with handles, these being connected by simple cranks to the axles. The four handles were each operated by a native using an up-and-down, or pumping action. A serious hazard was that the handles could give you a nasty crack on the head if you got in their way. These manual trolleys had no timetable so, if you heard a train approaching from front or rear, you had to stop and lift the trolley off the rails to let it pass. The motor trolley was an altogether different, more sophisticated vehicle that was built by Shelvoke & Drury and its use had to be integrated with other traffic. Then there was a caboose, or special coach, with living accommodation for use by my father or visiting VIPs.

This could be trailed behind any train and left for prolonged periods in a siding. With the fun we boys had in riding on these vehicles, it was inevitable that we should want one of our own. So we made 'go-carts', literally from a wooden box and pram wheels, and many a spill we had from these with permanently grazed knees, arms, and legs; much to my mother's concern. I have no doubt it was these early experiences on wheels that later created in me a love of the motor car which has lasted all my life.

Motors were very scarce in war-time Rhodesia so the family's only means of transport was a mule cart. Daisy the mule had previously hauled her owner up the main street, stopping at every bar on the way. This routine was deeply ingrained in Daisy who practised it with her new owner. Mother had to shuffle her feet on the buckboard, pretending to get out, before Daisy would start again and trot on to the next bar.

Father took us with him on some of his trips. Once we visited the British-owned Wankie Colliery, on the line towards Victoria Falls, which supplied all the fuel for the railway. The lumps and slabs of coal came up to the surface from the shallow mine on a wide inclined ramp. On another occasion we went in the caboose to Deroy, a settlement half-way between Umtali and Beira where the trains could pass each other by diverting to the long siding. The three of us set off into the jungle accompanied by natives carrying luggage on their heads. After two hours walking we came to a clearing with a circle of grass huts. Here was the headquarters of the Deroy mining enterprise! Father, ever hopeful, had joined with Brown and Murray, two prospectors who had found a reef of tin, to form the Palbroma Syndicate. We must have stayed nearly a week. I remember the huge fire built up each evening to keep the hunting animals away, but we still heard the calls of leopards and lions in the quiet of the night. Unfortunately, there was arsenic in the tin and there was just no economic way of separating the two. So the enterprise failed.

At the end of the war, with its inevitable reorganisation and retirements, Father was promoted to District Engineer in charge of the busy line between Salisbury and the Belgian Congo border; it being cheaper to ship out the copper from the Katanga Mines through Beira. We had to move to Bulawayo, and it was here he acquired his first motor car, a little-known make called a Saxon. This was a small American product (not more than 1.5-litres) with a two-seater body and dickey seat. It was not well suited to Rhodesian conditions. A sojourn of a year at Cape Town whilst father was moving, followed by several months long leave in England, opened my youthful eyes to the wider world of motor cars and must have determined the profession I was to follow. Back in Rhodesia, I kept in touch with that wider world by subscribing to *The Autocar*, *The Motor* and *Automobile Engineer*. I also wrote off for the brochures of different makes of car.

Auto-Architect

It was at Bulawayo that Mother had a period of ill-health. I already spoke the Mashona language and got on well with two of our servants, Lazarus and Sixpence, but Father decided that we needed a mother's help. Thus a formidable, no-nonsense widow came into our lives. Mrs. Hodgson's husband had escaped from unemployment in Northumberland and then died after working in the Northern Rhodesian mines. Her daughter married a man called Smith and their son, Ian, is not unknown to the world. As Rhodesian Prime Minister in 1965 he led the Unilateral Declaration of Independence from Britain (U.D.I.).

A Model T Ford was acquired for family transport, a car which I got to know from crank handle to back wheel (thanks to R.J. Nicholson's *Bible of the Ford*). This was followed by a Fiat 501, on the arrival of which I seized the Model T Ford, ripped off its bodywork, and spent all my spare time in building on it a rakish, two-seater sports body, plywood panelled, and inspired by an exotic Scripps Booth with a boat-tailed body owned by a local enthusiast. This was my first essay in motor body design and construction and it was not looked on with favour by my schoolmasters, who said it detracted from my studies. This was perfectly true. I failed three times in French and was then helped by a Swiss Catholic priest. Together we translated the Bible into French and this did the trick. Eventually I passed my London Matriculation Exams and the question then arose as to what I was to do. Father had paid for my brother to go to Warwick School, but could not afford the fees for a university education. I, anyway, preferred to

(Above and page 12) The Family Model T Ford once I had spent all my spare time building a plywood panelled two-seater sports body for it, inspired by the boat-tailed Scripps Booth owned by a local enthusiast.

11

Auto-Architect

get my hands dirty and had plans to drive the Ford on dirt roads from Bulawayo across the Limpopo River to Johannesburg. Father vetoed such an expedition (wisely after I had had the fun of making preparations!) and was keen that I embarked on an architectural apprenticeship as he saw a great future for that profession in developing South Africa. My interest was still in the motor car. I made a half-hearted attempt to get an apprenticeship with Bentley, then Father agreed a compromise that I should be apprenticed to a commercial vehicle firm as he also saw a future for African road transport.

In 1927 the Union Castle liner brought me to Southampton, accompanied by a large wooden box which my father had made to carry all those magazines and brochures. I was struck by the contrast between the bush and the rows of houses backing onto the track as we travelled to Waterloo. I went on to stay with an aunt at Whitby, and looked forward to the chance of some swimming, but the North Sea off the Yorkshire coast was just too cold after the African climate.

Chapter Two

Scammell and O.D. North

I was able to live with my Pierson Uncle and Aunt (she was my mother's sister) at Northwood in North West London. My immediate priority was to seek a suitable apprenticeship scheme, so I joined the Institution of Automobile Engineers (I.A.E.) as a student member. The secretary was extremely helpful and suggested Scammell Lorries as being a suitable firm for my circumstances. Situated in Watford, it was close to Northwood and I could get up to London for evening classes. I had an interview with the firm, they accepted me to join five other apprentices and, in the autumn of 1927, I commenced five years of unremitting and enthusiastic hard work. The journey involved just three stations on the Metropolitan Line and I would clock in by 8.30am. After a day assembling axles and gearboxes, or acting as mate to the engine-builders, I'd leave half an hour early to get back for some food and change into decent clothes. Then it was up to the Polytechnic for two hours in class on my degree course and back home by 11.00pm.

Scammell was one of the smaller vehicle builders and during the early 1920s they had produced articulated liquid tanker vehicles, mainly for the petroleum industry. Scammell had built, and was still building, its tractor units that were based originally on the American Knox design, popular for forestry work; but were beginning to emerge as one of the foremost constructors of specialist vehicles of limited production. This was largely because of the influence of their newly acquired, brilliant Chief Designer, O.D. North. For a young man like myself, who was totally interested in the design side of the industry, a more appropriate training ground could not have been chosen. For the whole period of my apprenticeship, I was privileged in a lowly capacity to work for this extraordinarily talented designer and to observe the working of his mind. He captured my total admiration and implanted in my mind, in no small measure, how to assess and solve design problems – 'How would O.D.N. do it?' one would ask.

Few people today have heard of this remarkable English designer, let alone of the products of his fertile innovative brain. In a long experience of the automotive industry, during which I came in contact with, and knew of, the work of the outstanding figures of the time, I knew of few who I could say were his equal. His products ranged from the light passenger car to the heaviest commercial vehicle, and all bore the stamp of his versatile imagination and genius. He has never been given the acclaim and credit he deserves, and I have no hesitation in singing his praises. He was active before the 1914 War, when he designed the crankshaft counterweight system on the Tourist Trophy and Grand Prix Vauxhalls, but came into prominence in 1922 when he designed the astonishing North Lucas Radial light car, only one of which was built. It was the most brilliant and original concept of light passenger car design of its day, perhaps of all time, and was North's idea of the future car. He realised that the basic foundation of the vehicle should be a strong, stiff box with the road wheel suspension medium necessary to traverse rough roads in comfort mounted relatively flexibly. The box of the NLR was a composite structure of aluminium, steel and wood; the pressed steel spot-welded techniques having not been developed in 1922. The four road wheels were carried on forged-steel arms which were independently pivoted transversely to the car's centre axis, the pivots containing the coil-spring suspension and an advanced form of hydraulic orifice damping.

The North Lucas Radial in Blackheath when it was probably about five years old. By this time the car had flared wings attached to the body.

Auto-Architect

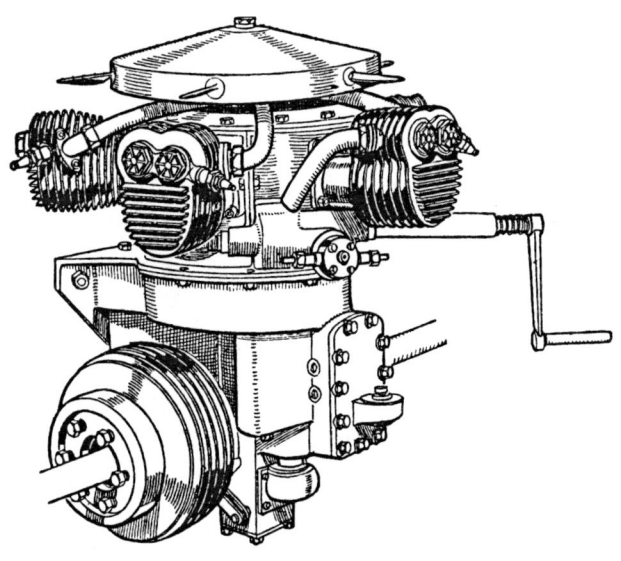

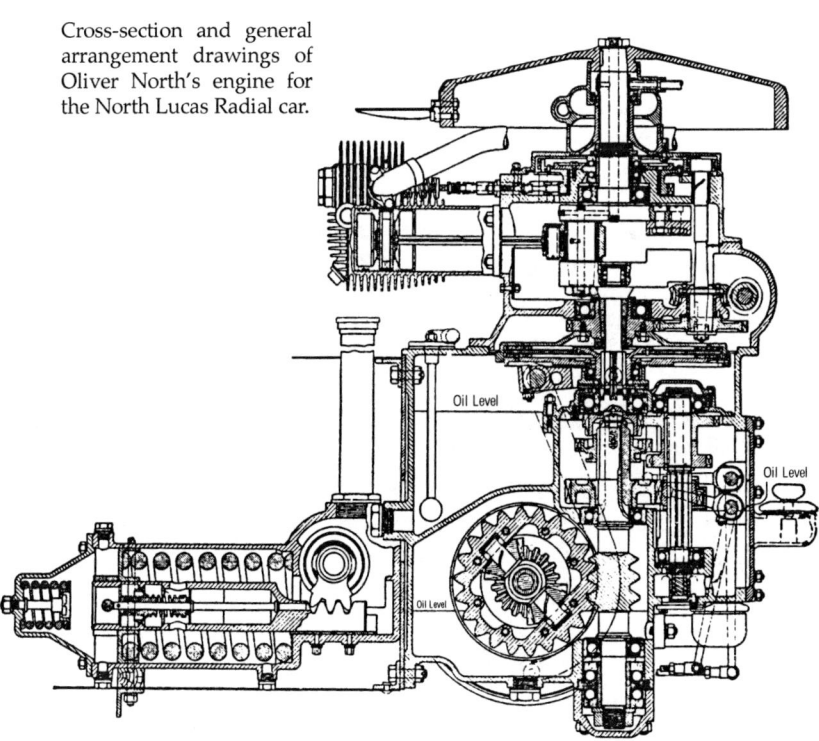

Cross-section and general arrangement drawings of Oliver North's engine for the North Lucas Radial car.

Auto-Architect

It was in the power unit and its location that the NLR departed most from the orthodoxy of its day. North was a strong advocate of the rear mounted position with its attendant use of air cooling, so he therefore adopted a five-cylinder radial arrangement with the crankshaft axis being vertical. Because of the larger diameter which would result from any overhead valve configuration, side valves were used. The original design overheated, but this was solved by changing to J.A.P. motor cycle cylinders and alloy pistons. This engine was installed above the clutch, three-speed gearbox and worm gear final drive, making a compact power unit which was cantilever-mounted from the rear transverse bulkhead of the box frame. Drive to the rear wheels was by universally-jointed half shafts. The NLR was certainly a design tour de force and received a great deal of detailed publicity in the technical press. However, it failed to find a sponsor, the reason I think being that it was too advanced and unorthodox in appearance. Curiously, the only production car which resembled it, the designer of which must have known of and been influenced by the NLR, was the Volkswagen. This is evidenced by a

---

ANOTHER VIEW OF THE NORTH LUCAS CAR

Edward Riddle, who became Chief Engineer at Scammell in the late 60s, and whose father was a cousin of Ralph Lucas, also recalls the car. He remembers Ralph's long-suffering wife, Mary, having to cope with a draughtsman from Scammell living in the Blackheath house for six months. Then in the evenings, Ralph, whose business since 1918 had been with Thames lighters, would pore over the drawings. Oliver North came once or twice a week.

When it came to travelling in the completed machine he was relegated, as a youngster, to the back seat and much more aware of the noise and vibration of the engine. He also remembers the light interior resulting from the transparent roof made from doped canvas following aeroplane practice of the time. Oversteer was an initial problem mitigated by suspension fine tuning.

On the debit side the family were inclined to comment that Cousin Ralph could have got himself a couple of Rolls-Royces for the cash consumed by the project, but he obviously enjoyed the kudos that came from showing the car to other manufacturers. And, if in the end none of them had been confident enough about the sales prospects of such a different design to commit to production, Ralph himself had nearly 70,000 miles enjoyable use of the machine. In the post-war years Oliver North is on record saying they had attempted too many innovative features at the same time.

patent of Ferdinand Porsche (No. 431737 of 1932) which covers the mounting of a five-cylinder, air-cooled radial engine at the rear of the fast back body, the only novelty being that the crankshaft axis is at right angles to the external body shape, instead of being vertical as in the NLR. Presumably Porsche found there were difficulties in the five-cylinder radial arrangement, particularly with an overhead-valve cylinder head, so opted for the more easily accommodated four-cylinder, opposed arrangement. The general layout and specification of both of the two cars are strikingly similar.

Ralph Lucas, previously responsible for the Valveless car, had arranged the finance for the project from his Uncle Frank and he regularly used the prototype after the initial testing. It was still in use when I was at Watford and I made friends with a senior mechanic called Crowe, who did the servicing. He sometimes took me with him when checking the car on the road. The day he allowed me to drive is still clear in my mind with the abiding impression of a silent, sweet machine, the suspension cushioning the bumps and the noise behind and remote as if in another vehicle, but there was some roll on corners. When Lucas finished with the NLR it was sold to another apprentice called Farquharson. We met again at a Shell reception after the war and, when I enquired after the car, he told me that it had been broken up. It should have gone to the Science Museum. I wish I still had the energy to reconstruct the NLR from the drawings published in the Automobile Engineer in 1922.

A six-wheel-drive Scammell chassis with a driven front axle was developed for military use; this had the ability to scale short sections of vertical wall.

Auto-Architect

(Left, below and opposite) Drawings of the ingenious four-wheel drive rear axle used by Scammell. This employed one worm drive and differential unit to power all four wheels, using separate gear trains housed in casings on the ends of the rigid axle. These casings were free to rotate to maintain all the wheels in contact with the ground over rough terrain.

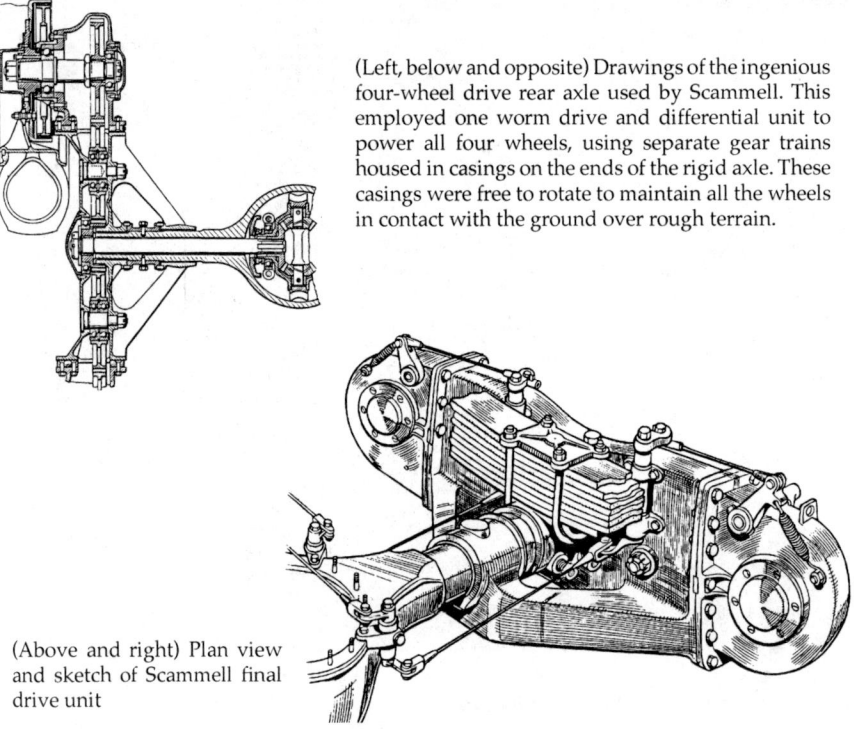

(Above and right) Plan view and sketch of Scammell final drive unit

18

Auto-Architect

The seven-ton, six-wheel Scammell had remarkable cross-country ability. The single rear axle carried pivoted cast housings, each of which contained a train of spur gears to power the rear wheels. The un-powered front axle was supported by a transverse leaf spring that was pivoted at the centre.

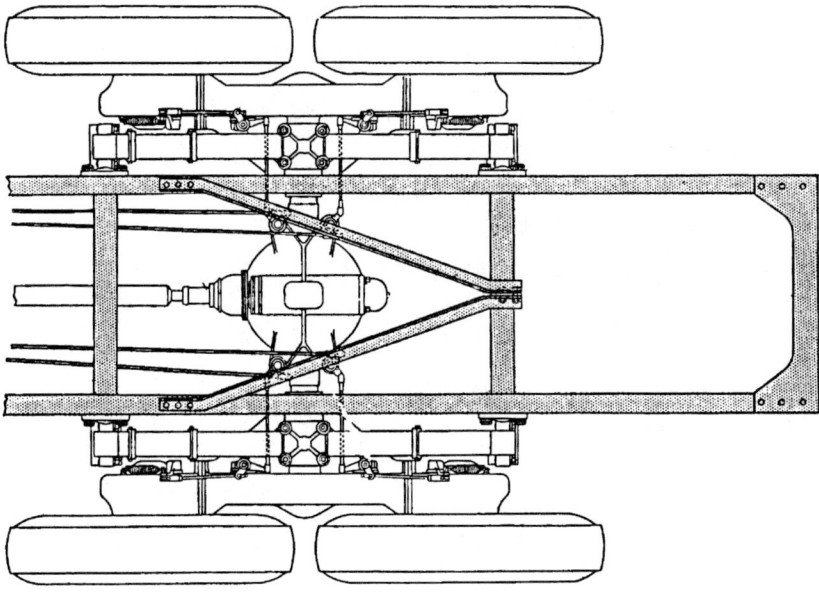

General arrangement drawing of the rear of the six-wheel Scammell chassis.

Auto-Architect

The influence of North was to be paramount at Scammell. Largely due to him the company was just entering a period of remarkable innovative design in the commercial vehicle field. It had been kept afloat during the depression by repeated orders for articulated road tankers for the major oil companies. A need had arisen for a transporter to carry large steel pipes for the oil pipeline from the Gulf to Haifa over very rough desert terrain – it had to possess outstanding off-road capability. North produced the remarkable cross-country, seven-ton, six-wheeler. Its cardinal feature was a single worm-driven rear axle mounted on which, on both left and right, were massive pivoted, articulated cast housings. Each contained a train of spur gears which drove stub axles which carried the road wheels. The simple front axle was supported by a transverse spring pivoted at its centre to the frame. This arrangement allowed exceptional vertical rise and fall of each road wheel, about plus or minus one foot, so that the vehicle's driveability over rough terrain was remarkable. It even became apparent that it had military applications; a six-wheel drive version was produced with a driven front axle. This had even greater off-road capability, being able to climb a vertical wall! The first versions were used as gun tractors, but then became widely used as tank transporters in World War II. Concurrently, a need developed for a special

The remarkable cross-country ability of the seven-ton, six-wheel Scammell was not confined to military use as there were many civilian applications for the innovative technology employed in its design.

Auto-Architect

North designed for Scammell this vast, 100-ton, transporter. It was so long that the rear wheels had to be steered by an operator at the back who was in telephonic communication with the driver.

Auto-Architect

vehicle to haul payloads of one hundred tons on the public highway. Scammell was again approached and the genius of North produced a solid-tyred, fourteen-wheel leviathan. It was articulated with four driving wheels on the tractor and a bogey of eight wheels at the rear of the swan-neck trailer. This was so long that to negotiate main road bends the bogey had to be steered, there being telephonic communication between the steersman and the driver in the tractor. This monster was propelled by a five-litre, four-cylinder petrol engine, bottom gear ratio was in the region of 500:1!

Between these major projects North's versatile mind was exercised on other tasks. One of the most important was the complete redesign of the tractor-trailer unit known as the Mechanical Horse. It was a replacement for the horse drawn drays used by the railway companies to distribute merchandise from their terminals in congested city centres. The project had been initiated by the Napier Company, but as they did not have sufficient capacity with their aero engine work it was handed over to Scammell. North abandoned the four-wheel tractor and replaced it with a three-wheel unit, with a single wheel at the front so that it could literally turn in its own length. He also designed an automatic-coupling device, so that the tractor could be reversed up to the trailer and either connected, or disconnected, via controls situated entirely within the cab. It was easily the most manoeuvrable and easily operated vehicle for city centre use in its day.

With the adoption of the six-wheel drive vehicle for use as a gun tractor, it became apparent that the old Scammell four-speed gearbox was inadequate.

O.D. North completely redesigned the Scammell 'Mechanical Horse' tractor/trailer units, abandoning the previous four-wheel tractor in favour of a three-wheel arrangement giving them the manoeuvrability required for local delivery of goods sent by rail.

Auto-Architect

An up-to-date gearbox with six speeds was needed. Here again North eschewed the conventional approach and produced a compact unit having only four spur gear trains, the two extra ratios being obtained by compounding; all gears being engaged by dog clutches. The gear selection mechanism was complicated and qualified for the old adage *C'est brutal mais ça marche!*

I saw North through the eyes of a youthful apprentice who had aspirations in the design field himself. With all this innovative work proceeding through the drawing office and the workshops, my admiration for him was unbounded. He had the ability to reach the nub of a design problem immediately, and then produce the right solution. In a subsequent stage of my career, I had the privilege of meeting and talking to Dr. Fred Lanchester. I was immediately struck by the likeness of these two men. They both had the same penetrating mind and ability to solve problems.

Having finished the practical side of my apprenticeship in the works at Watford, I was transferred to the design and drawing office which, oddly, was in Holborn, London. This was advantageous for me as I was able to walk from office to evening classes to complete my degree course. London was an exciting and interesting city to be in at that time, with a lot of activity in the sciences, arts and politics. I had become a graduate of my own professional body, the I.A.E., and was able fully to take part in its activities and meetings. They were good venues for associating with people of like interests. Before long, I found myself elected Secretary of the London Graduates Branch, and finally Chairman, in which capacity I had to present a learned address. I chose as my subject *The Control of the Infinitely Variable Gearbox*. There was, at that time, a great interest in infinitely variable gears culminating in the introduction by Austin of the Hayes gearbox on one of their models. The control system used left something to be desired, so I evolved a theoretical system taking engine characteristics into account. Horsepower, being a function of torque multiplied by engine speed, can be generated in an infinite number of ways depending on the stable range of speed at which the engine can run. Thus, a given horse power can be produced by a given torque multiplied by a given engine speed. It can also be produced by half that torque multiplied by twice that speed, or by an infinite number of speeds within the stable range. But there is only one speed for minimum fuel consumption in this range. Therefore, to minimise consumption, the engine should be at that speed at all road speeds. This is only possible with an infinitely variable transmission which, by its engine speed control, can produce fuel consumption savings of up to 35%. Although the system described in my address in 1934 is a mechanical one, it anticipated in principle the sophisticated electronic systems used in the present day infinitely variable transmissions. I tried to patent it but found that General Motors Research in the U.S.A. had been granted a patent identical in principle about a year earlier.

Auto-Architect

There was another consequence of the fortunate choice of Scammell. A Miss Diana Varley was working in the drawing office. She was that, then, rare species, a girl who was planning to become an engineer. The grand-daughter of S.A. Varley, who had worked with Michael Faraday, the inventor of series and compound dynamo winding, she had been educated at Oxford High School. Her training was sponsored by Marryat & Scott, the London lift engineering firm (the directors were family friends). The spell in the Holborn office was part of her work experience. We soon found we had ideas and aspirations in common. In the summer of 1938, defying the convention of the time, we travelled together by train to the South of France and visited Sainte Maxime and Saint Raphael. It seemed, as many young couples have since found, the natural and sensible way to get to know each other. We continued to get on well and were married in the Oxford Registry Office on 6th May 1939.

With Diana at the Registry Office, St. Giles, Oxford on our wedding day, 6th May 1939.

24

Chapter Three

The Deroy Adventure

Although my five years apprenticeship was in the commercial vehicle field, and as an engineering design training could scarcely have been better, my interest and ambition was in passenger cars, more particularly sports and racing cars. Any spare time was devoted to special building and to experimenting with suspension ideas. Uncle Henry and Aunt Sarah were very forbearing and encouraged me to do my own thing in their garden shed. After my African experience my thinking was already channelled towards the replacement of leaf springs, thus achieving a more comfortable ride to cope with those rough surfaces. For my first effort I used bungee rubber blocks which were then a De Havilland component for their light aircraft undercarriages. The Morris Cowley rear axle was controlled by radius arms. A Roesch-designed Talbot engine and Citroen front axle completed the package. It wasn't finished (though I later sold it to a chap from South London for £10) because another I.A.E. graduate, Chalenor Barson (also remembered for his various Barson specials), asked me to design a new sports car ordered by a wealthy friend, Joan Richmond. She had already won races at Brooklands and Donnington Park.

A more naive couple of young men could scarcely have been found! We firmly believed we could make a success of the project. However, Joan then decided to return to her native Australia and we lost our cash base. Luckily, I was in consultation at that time with a patent agent, F.J. Cleveland, to whom I had been introduced by Diana, to patent features of my suspension system. He was in touch with Antony G.A. Fisher, just down from Cambridge after reading engineering, who was looking for experience in setting up small businesses. Our project appealed to him, so he agreed to provide £1,000, and we set up a company. A garage was rented in Penge, and Chalenor worked full time skilfully constructing a car from my detailed drawings.

Whilst I was still working in the Scammell drawing office, I went down once a week. We submitted all expenses direct to Fisher who kept the books. We decided to name the car the DEROY after the location of my father's Palbroma tin mine in Mozambique - it seemed an appropriate name for a new, fast and svelte

Auto-Architect

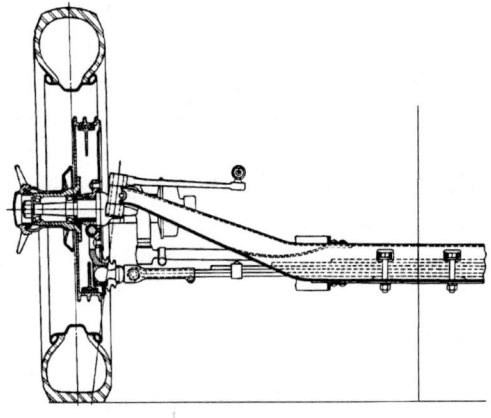

(Above) The rear suspension of the 1937 Deroy sports car. The springing was by transverse torsion bars set across the car and the vee-shaped axle beam acted as an anti-roll bar.

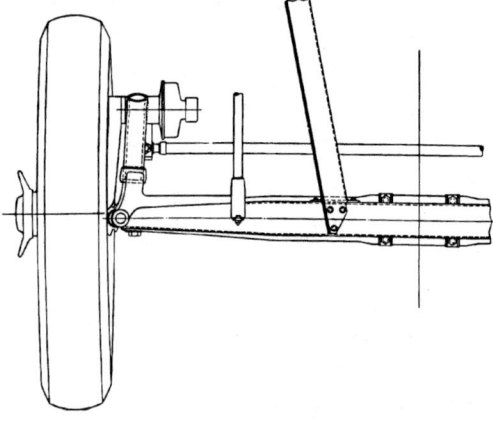

(Left and opposite) My drawings of the front and rear suspension of the Deroy published by the Institution of Automobile Engineers.

sports car; and so the Deroy Car Company began its short career. The prototype was completed and evinced considerable admiration from those to whom it was shown. Its styling was praised as being equal to that of many of the continental trend setters of the day. Except for the power unit, its mechanical specification was advanced; the front suspension was independent, of modified Dubonnet type using a transverse leaf spring. The rear suspension, also independent, might be said to be of the De Dion type, to give very low unsprung weight, but in which the vee-shaped axle beam was used as an anti-roll bar - a feature which predated its use on post-war Saabs, Panhards and others. The springing was by transverse torsion bars set across the car just ahead of the beam.

Dr. F.W. Lanchester noted my efforts when, in response to his appeal for examples of independent rear suspension for publication in an I.A.E. paper, I submitted my design which he accepted and praised. He seemed pleased that a British designer had ideas which matched up to European thinking. This was a considerable boost to my confidence. Another novel feature was the hood, with the canvas held taut by a metal panel which was extended on struts and then acted as a cover when lowered. The mistake I made was to use a side-valve, four-cylinder 1.5-litre engine which, despite the use of twin carburettors, was woefully lacking in power so that the performance of the car was quite inadequate. This was an engine made by Scammell for their three-ton Mechanical Horse and which I could buy for a reasonable price, but it was a bad decision and ultimately lead to the project being abandoned. After I had left Scammell in 1935, concerted efforts were made to interest people like Charles Hurlock at AC, and R.G. Sutherland at Aston Martin but they were not prepared to adopt the design. It was obvious that a

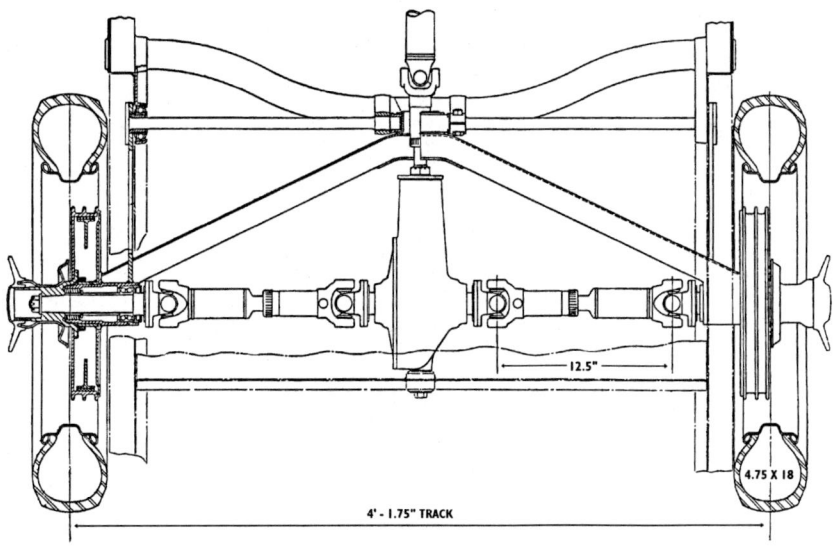

Auto-Architect

The Deroy hood packed away neatly behind the seat and was then concealed by a metal cover. In a totally original manner, this cover also formed the support for the hood fabric when this was erected.

more powerful proprietary engine was required, and a great deal of development needed to be undertaken. This meant a lot more cash. With the ominous shadow of a European war looming in the future, no one (including Fisher) was prepared to risk more capital on such a project, and we closed the company in 1938. I was given the prototype car in compensation. Chalenor emigrated to South Africa where he had been offered work by Reckitt & Coleman erecting steel-framed factory buildings. It was a pleasure to meet again at Hout Bay in the 1960s when I visited my sister in Cape Town. I was then at Vauxhall and was able to travel around that beautiful country in a car lent by my employers, General Motors.

I remain indebted to the man who gave us this opportunity. The British motor industry attracted, but often failed to retain, well-educated, independent individuals. Sir Antony Fisher (he was knighted shortly before his death in 1989) went on to found the Buxted Chicken Company (highly successful at the time whatever the view of battery farming today) and then the Institute of Economic Affairs. In his memorial address Lord Harris, Chairman of the I.E.A., referred to Antony's belief that 'independent scholarly publications would move men of goodwill to topple the false god of unlimited government'.

The Deroy with the front wings removed to reveal the independent Dubonnet-type front suspension, steering, and hydraulic brakes. Given a more suitable power plant the light weight and low frontal area would have ensured good performance. My mistake was to opt for the Scammell engine to save cost.

Auto-Architect

THE DEROY SUSPENSION

In October 1997 Chalenor Barson, then 88, wrote from Hout Bay, South Africa: 'I drove the Deroy quite a lot testing the suspension and making it work hard driving over curbs and things like that. I once went over a rather sharp humped-back bridge and I, or rather my back-side, had problems with bottoming through the seat onto something hard! That was one of the problems. It was a bit too soft and that was a very painful realisation. I think it was a good suspension both back and front which would take quite large variations in service. If it was a little on the soft side, it still suited the car.'

There have been many articles over the years in praise of the prewar B.M.W. sports cars. In the Automobile, December 1997, Doug Blain wrote of the chassis of the 328 'which made it such a unique precursor of things to come. It was the rigidity of that chassis, rather than its springs; the delicate balance between spring rates, wheel movements and shock absorber settings that gave the handling such consistency and the ride such suppleness. These were revolutionary concepts in 1937.' Yes; but it was not unique. If the potential customers had but known, Gerald's design was waiting in the wings, although it needed further development and a thoroughbred power unit.

Chapter Four

Nuffield and the Oxford Vaporiser

My immediate concern was to get back into the main stream of the industry and it was here that my membership of the I.A.E. was invaluable. Through the good offices of the secretary, Mr. Brian Robbins, I was able to get an interview with Mr. Cecil Kimber, Managing Director of the M.G. Car Company. I went to Abingdon to meet him in the Deroy. He took a long time examining it and asking about its main features, and was obviously impressed. He then told me that all design and drawing office work had been transferred to the main Morris office at Cowley, so he immediately phoned Mr. A.V. Oak, the Morris Technical Director, and arranged an interview for me. This proved successful and in due course I was offered the job of being in charge of M.G. work in the Morris Drawing Office. This was just the opportunity I was seeking and was readily accepted.

My first assignment at Cowley was to work on the Series YA, a small, close-coupled saloon utilising Morris Series-E side panels, doors, trunk lid, and the 1¼-litre XPAG engine. It had been designed with a box-section frame and independent front suspension. This was proving expensive, and an alternative, conventional arrangement was needed for cost and performance comparison. This was comprised of an orthodox beam axle with half-elliptic leaf springs, shackled at the front to a channel-section frame. An anti-roll bar was carefully positioned to absorb the brake torque and reaction. I incorporated proprietary cam and lever steering. Due to the onset of war in 1939, this design was never used and car work was put into cold storage. When this M.G. saloon was eventually announced in May 1947, the design had reverted to coil-sprung, independent suspension with the frame slung beneath the rear axle. There was a panhard-type rod (then known as an anti-sway bar) connecting one end of the axle to the other side of the frame. It was a design with above average handling for its time. I did not, as has been suggested by other writers, design the body. This had been done in the Morris drawing office as an extension of Leslie Hall's design for the Morris Eight saloon.

Once war broke out everything seemed to stop, but before long, through the grapevine, it became known that we were to set up a production line for Tiger Moth training planes as all the De Havilland factory space at Hatfield was needed for production of the all-wood Mosquito. The engines were supplied by the parent company, and the planes were to be flown off from an airfield under construction adjacent to the Morris works. In this set-up I was assigned the job of producing a parts list of the plane, necessary but boring, and requiring considerable liaison with De Havilland resident staff. By the end of the contract, Cowley had put together over 3,000 Moths, a significant contribution to the many thousands produced throughout the world. With the increased tempo of the war on the Western Front, particularly in the air, Morris Motors became one of the Civilian Repair Units with responsibility for repairing fighter aircraft damaged in combat. Spitfire and Hurricane aircraft, less their power units, were delivered to the works on low loader trailers and, after careful assessment, detailed repair schemes were prepared. The damage usually consisted of gaping holes in the fuselage or wings, caused by cannon fire, and in bad cases the proposed repair had to be approved by the parent company before proceeding. Many aircraft were restored to service by this and other units during the critical 1940 Battle of Britain period, contributing in no small measure to our superiority in the air. I also played my part as a humble private in the Factory Home Guard and once a week I was assigned to all-night fire watching duties at the new airfield. I was often in the company of a veteran soldier who regaled me with his memories of the Battle of Omdurman; the roar

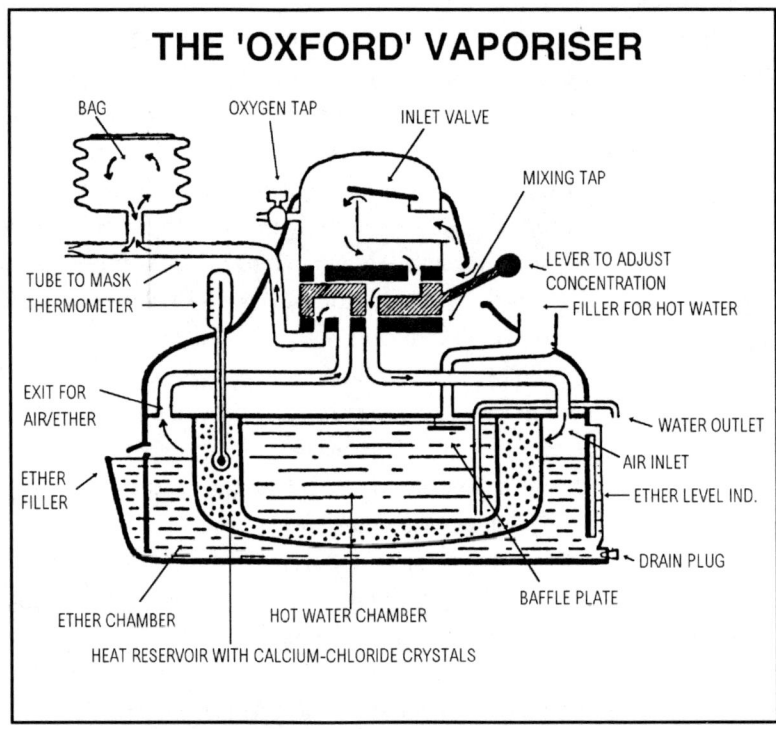

Auto-Architect

of 60,000 advancing Dervishes cheering for the Prophet and the Holy Khalifa, the colour of their white and yellow standards, the vicious fighting when sword and spear met pistol, and the hacked and mutilated bodies when it was over. In those long watches before dawn the stories seemed to relate to such distant times. In fact, Omdurman, which led to Kitchener regaining Khartoum, was fought on 2nd September 1898. (Churchill described his part in the Cavalry Charge by the 21st Lancers in his book *My Early Life*.) If my old soldier had then been 19, he would only have been 62 in 1941. In the morning we went straight back to work. We avoided all the bombs but at least the lack of sleep helped us to feel that we were not shirking. Alec Issigonis, by then higher up the ladder, was one of those who slept in the offices ready to be called if needed.

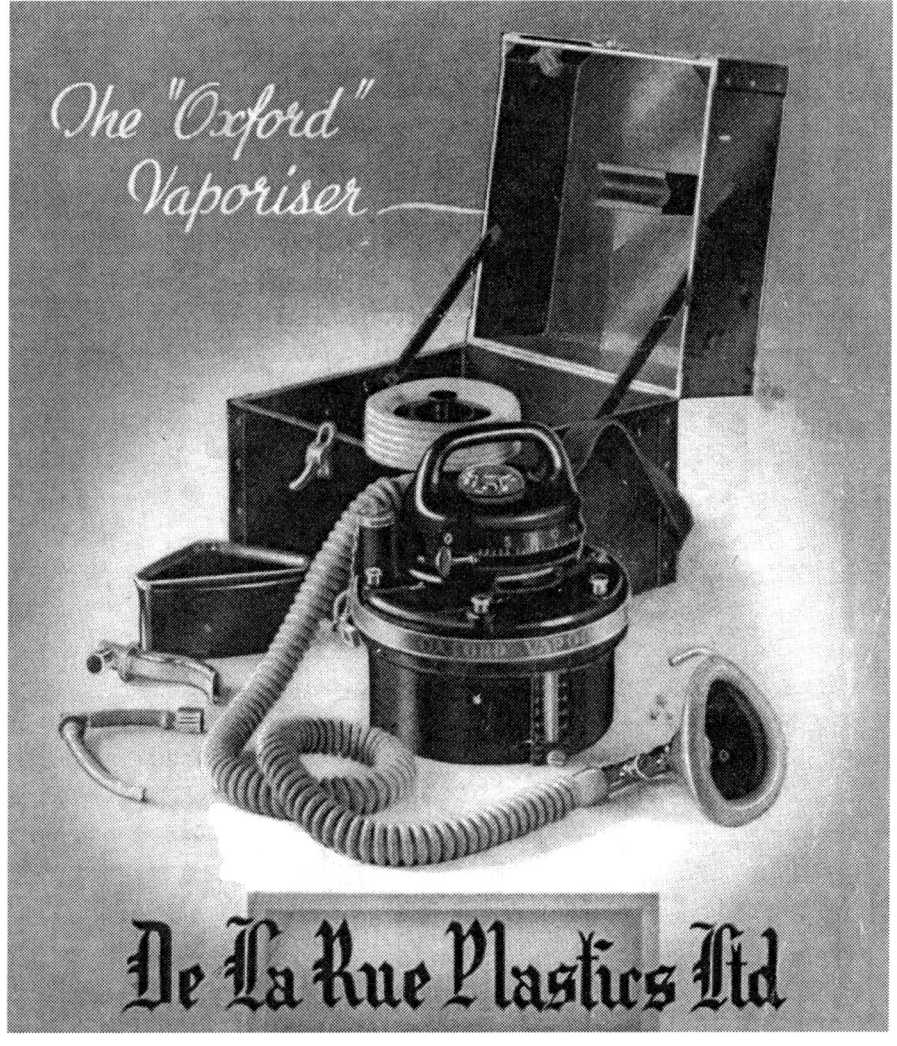

Auto-Architect

After working on aircraft repairs for several months, and in a most unexpected and welcome way, I was chosen to undertake an interesting original design project. I was summoned one morning to go to the office of my chief, Mr. A.V. Oak, there to meet Robert Macintosh who was Nuffield Professor of Anaesthesia at Oxford University. On Oak's desk was a device which at first glance looked like a boiler for a model steam engine. It was a fabricated brass cylinder about ten inches in diameter and six inches long, with a lever protruding from one end. I was told it was the basis of a portable anaesthetic apparatus which enabled surgical operations to be performed in hazardous situations in warfare where field hospitals were not available. It could be carried and set up by medical orderlies. I later learned that the idea had come to Macintosh while observing medical arrangements in the Spanish Civil War, and had been developed on his return by his staff in Oxford (Messrs. Epstein and Suffolk). Lord Nuffield, with his usual beneficence, had agreed to fund the project, to set up a production line, and to present an undisclosed quantity to the three branches (Army, Navy and Air Force) of the Royal Medical Corps. The trouble was that while the crude device on Oak's desk worked, it was quite impractical to produce in quantity - which is why he called me in to carry out a complete redesign.

The apparatus basically consisted of three concentric cylindrical chambers, an inner one containing hot water, an intermediate containing calcium chloride - which had the characteristic of absorbing heat rapidly and dissipating it at a slow uniform rate, and an outer containing ether. In operation hot water from a portable heater was poured into the inner chamber, the heat being rapidly absorbed by the salt in the intermediate chamber and then passed more slowly, and at a uniform rate, to the ether. This vapourised and was mixed with air before being passed to the patient. The strength of the charge going to the patient (the ether/air ratio) could be controlled by the manually operated mixing valve. A delicate disc valve was provided in the inlet stream to monitor a patient's breathing - if it stopped, they were dead.

The project was overshadowed by the wartime restriction on the use of strategic materials, some of which were needed. I was allocated one pound of aluminium, eight ounces of bronze, six ounces of stainless steel, but no limit on plastic materials. Furthermore, I could dip into Lord Nuffield's pocket as deep as was necessary, which in practice meant there was no restriction on sophisticated tooling. It was soon apparent to me that this was a design exercise in plastics, a field in which I had no experience. The first thing I did was to arrange to visit the De La Rue Company in London (the potential suppliers of plastic mouldings), to learn something of the materials I could use and the design techniques available. In reviewing the prototype design, which used a barrel-type mixing valve, I decided this must be abandoned as it entailed accurate machining operations to maintain a reliable and repeatable radial clearance and hence mixture strength; in fact selective assembly would be required which was not acceptable. I replaced it with a spring-loaded rotatable plate valve rather like a modern water tap; this could easily be lapped-in if the virgin surfaces of the plastic mouldings were not immaculate enough to give a leakproof condition. It also facilitated the graduation of the valve and gave greater consistency. This major change was entirely successful and having made it, the rest of the design was comparatively straightforward. It consisted of an open plastic bowl about ten inches in diameter and six inches high, the open end of which was closed by a thick 'sandwich' pair of plastic discs between which the ducting for the air and the ether vapour was formed. The disc was surmounted by the mixing valve. This redesigned vaporiser could be produced in quantity on an assembly line.

The biggest development problem was to establish a reliable gasket between the bowl and the discs - I learnt rather too late, as tooling was well

Although my war work kept me out of active service I 'did my bit' in the Home Guard, spending many long nights on fire-watch duties at the airfield adjacent to the factory.

under way, that it is better to have lots of small bolts rather than a few large ones. In the end Ciba-Geigy came to the rescue with a suitable adhesive to be used in conjunction with the gasket. An air-pump to stimulate the patient's breathing was needed and I designed this as a simple spring-loaded, bellows type which was mounted on the top disc. The vaporiser with its accessories, such as the face-mask, tube, spare ether bottle, etc., was housed in a stout wooden box which had a shoulder strap. These were made by Chappel Piano Company. Being fairly light, it could easily be manually transported.

With the final approval of the Oxford Vaporiser 'off tools' by Professor Macintosh and his team, and the setting up of a production line, my work on the project finished. Although I was never mentioned in connection with it, and did not expect to be, it was very satisfactory to have been so closely involved in a device which I believe saved many lives during the war, and which got considerable praise as an article of good industrial design. If I wasn't fighting at least I was helping those who were. The vaporiser was successfully used in the Falklands campaign and variants of the device continue to be developed.

After the feverish activity on the vaporiser there was a lull, although for a short time I worked on an odd project which was the brain child of John Morris, the brilliant chief engineer of the S.U. Carburettor Company. This was an opposed piston, two-stroke engine with a Scotch crank, an alternative way of converting reciprocating motion to rotary motion using two crankshafts and sliding blocks, instead of one orthodox crankshaft. It was not proceeded with. Morris was a fanatical advocate of front-wheel drive and used nothing but Traction Avant Citroens. Being a great friend of Alec Issigonis, it is my guess that he influenced him in adopting front-wheel drive in later BMC models.

Despite the fact that the country was engaged in a life and death conflict, one's thoughts occasionally turned to motor cars and their problems. So imagine my astonishment when I read in one of the technical periodicals an advertisement by a motor manufacturer for a chief designer. I could not guess which of the few it could be so out of curiosity, and with nothing to lose, I decided to find out by answering the advertisement. It turned out to be Jowett, a minor manufacturer in Bradford, and I had an invitation to go for an interview from the managing director, Charles Calcott Reilly (who always used his full name in correspondence).

Chapter Five

Designing the Javelin

Jowett was a very small concern relative to the big five manufacturers and had undoubtedly been losing ground in the highly competitive market of the 1930s. Had it not been for the advent of the war it might conceivably have gone under. The founders and original owners of the company were two practical engineers, Benjamin and William Jowett, who had designed and built a simple two-cylinder, horizontally-opposed petrol engine which they installed in a robust chassis with a rudimentary two-seat, open body.

This car they launched on the market in 1910 and, aided by war work from 1914 to 1918, it was produced profitably until 1939 – latterly with more elaborate four-seat coachwork. The two-cylinder engine, also used in the post-1945 Bradford van and utility, is credited with having had the longest production run in the industry, but it was improved and modified a great deal over that period. So, indeed, was the financial structure of the company itself for, due to falling profits, it was floated on the Stock Exchange in 1935 with a view to attracting additional capital for investment. This unfortunately it failed to do; so in a rapidly worsening situation a major board room shake-up had to be made, with three independent directors appointed. One of these was Reilly who subsequently, in 1939, was made managing director. He was an energetic, forceful and likeable character, and had been an associate of E.C. Gordon England, who produced small cars in the London area based on Austin Seven chassis. His first move was gradually to alter the factory layout to a flexible war footing, enabling a great variety of items to be produced - ranging from mortars, field guns, and ammunition to aircraft components and capstan lathes. At the same time, he obtained a large order from the Admiralty for generating sets fitted with the two-cylinder Jowett engine.

Having successfully geared up the firm for war production, he began with characteristic foresight to think and plan for the post-war period when returning servicemen would be seeking jobs. I visited Bradford and met Reilly and other members of the firm, with whom I was impressed, but the general environment, size and standing of the company was not attractive enough for me then to accept the position I was offered. I reflected that although its products would have sold for a time in a post-war, car-starved market, as it had no up-to-date body plant I could not see it competing in the longer term against the big companies. I could predict a future for it in a slightly higher price range, say £220-250 at pre-1940 prices. Even then it would need a product having some outstanding virtues and character, with high export potential, as Britain would have to export to pay her way in the world. However, although I had a sound position at Morris Motors, the prospect of having a free sheet of paper to design a new car was very enticing and would satisfy a long held ambition. The matter was settled by a surprise visit by Reilly to my home in Oxford, when he persuaded me to change my mind and accept his offer.

  I moved to Bradford in January 1942, my wife and one-year old daughter, Celia, following six months later. For them it was a traumatic move from a romantic, thatched cottage in Oxford to a third-floor flat in a grimy city, but Diana supported me loyally. I was allocated a small back room in the offices of the Jowett factory at Idle, no larger than a bathroom, with a drawing board and lay-out table. There I set out my ideas on a future Jowett which was to become, with no basic changes, the Javelin. My concept for the car was of a six-passenger family vehicle with reasonable luggage room, minimum overall size, above average performance, and economy. Low weight was vital; the target was 2,000lbs. Also important were good comfort and road holding, and the ability to traverse rough roads. My own experience suggested the need for eight inches of ground clearance. To achieve this economically, interior trim and appointments were to be simple and durable, certainly not de luxe. Low wind resistance was to be aimed at with the layout enabling a streamlined body of distinctive and different shape. There was a widely-held opinion that all passenger cars would ultimately look like a teardrop, or a fish. Experimental designs had been built on this theme

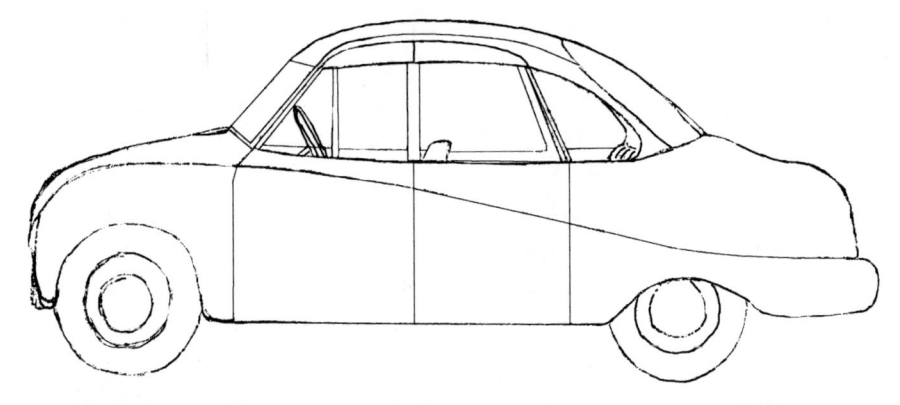

The Naum Gabo proposed design for the Javelin body envisaged using the basic mechanical arrangement I had designed. In the event it was rejected, partly because it would have been too costly to build.

and the one that caught my eye (and therefore presumably influenced me) was conceived by John Tjaarda for the American Briggs Company. It undoubtedly influenced the styling of the Lincoln Zephyr. In general my concept was what is now known as a world car and, in my opinion, was very much in harmony with the Jowett tradition and should be acceptable in world markets, like those I had known in Southern Africa.

The basic aim of car design at that time was to use the maximum part of the overall length for passenger accommodation and luggage, and the minimum for the power plant. To achieve this a compact engine was necessary, such as a flat or V-twin, or four. The Austrian Steyr and the German VW had already shown what was possible. Jowett's whole past had been linked to such units but their merit had not been fully exploited and they had been thinking of changing to the simpler in-line four, an example of which was at the factory. Steve Poole, thoroughly sound if old fashioned, was chief engineer. He was always welcoming and didn't try to influence me. Despite the fact that the horizontally opposed layout is inherently more expensive, this still seemed the opportune time to continue the Jowett tradition with a new overhead-valve unit to replace the pre-war sidevalve, coupled to the updating of the total concept. I showed Steve aspects of the design and listened to his suggestions. When he realised my ideas were fitted to the Company's assessment of the market, he gave, and continued to give, his full encouragement whilst he concentrated on the flat-twin Bradford, our passport for survival until the Javelin was ready for sale.

These ideas were put to Reilly - Peter as I was to come to know him. He was christened thus by his second wife, who didn't like Charles, and this name was used by most of his closer associates. However, other designs had also been investigated. Herbert Read had brought together a group of industrial designers and was offering help in styling post-war products. Peter, ever enthusiastic, had decided that we should send them a drawing of the basic seating plan and mechanical units. This was given to the Russian émigré designer, Naum Gabo, who produced a futuristic concept, complete with scale model, which was brought up to Bradford in 1945. I was concerned. I didn't want to lose control; but Peter accepted that the design, though interesting, was far too advanced. Amongst other things, there was the requirement for curved glass which was not then available.

I knew that O.D. North had also come to Idle and there had been discussion of the North Lucas Radial concept. Roy Fedden had visited with his ideas for a world car, another space-conscious design with, like the N.L.R., a compact radial at the rear. I was, therefore, relieved when Peter told me that he was in total agreement with my plan, a less complicated route to the same goal. We were all thinking along the same lines: North, Fedden, and Porsche,

Auto-Architect

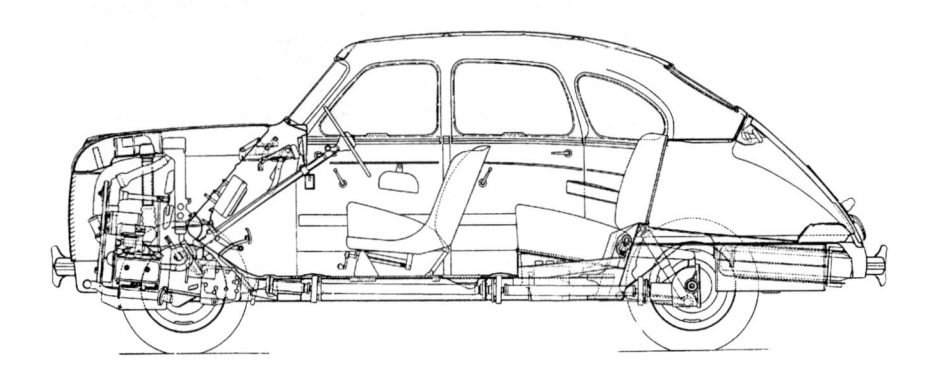

My drawings for the Javelin. Above is the side elevation general arrangement which clearly shows the smooth shape I was trying to achieve, whilst giving the maximum room to passengers and luggage, and below is a section through the 'flat four' engine. Opposite are the independent front suspension, the rear suspension and a drawing of the then-fashionable steering column gear lever mechanism.

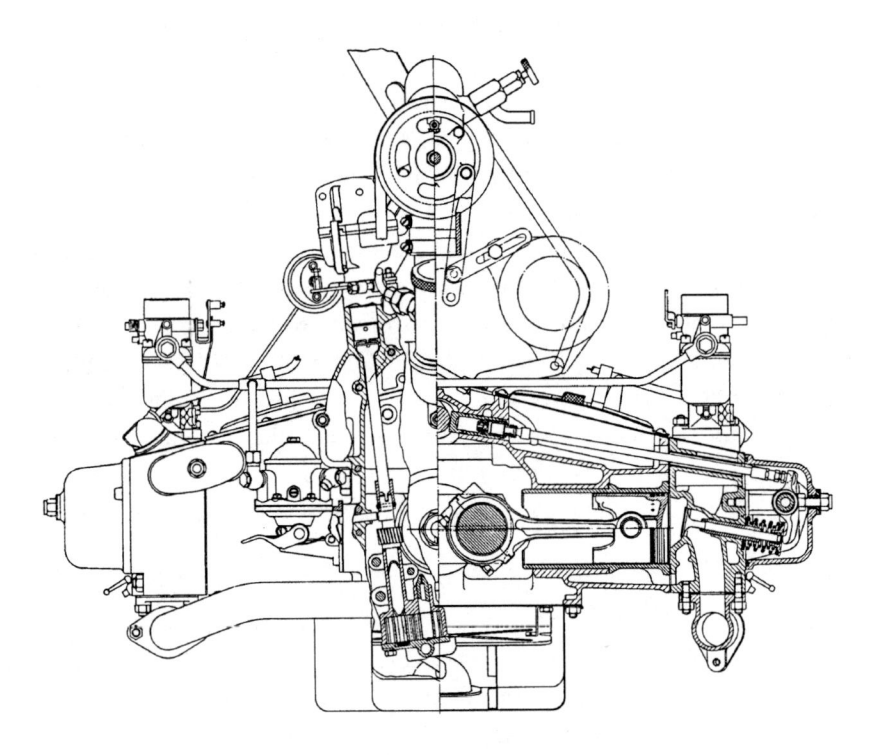

Auto-Architect

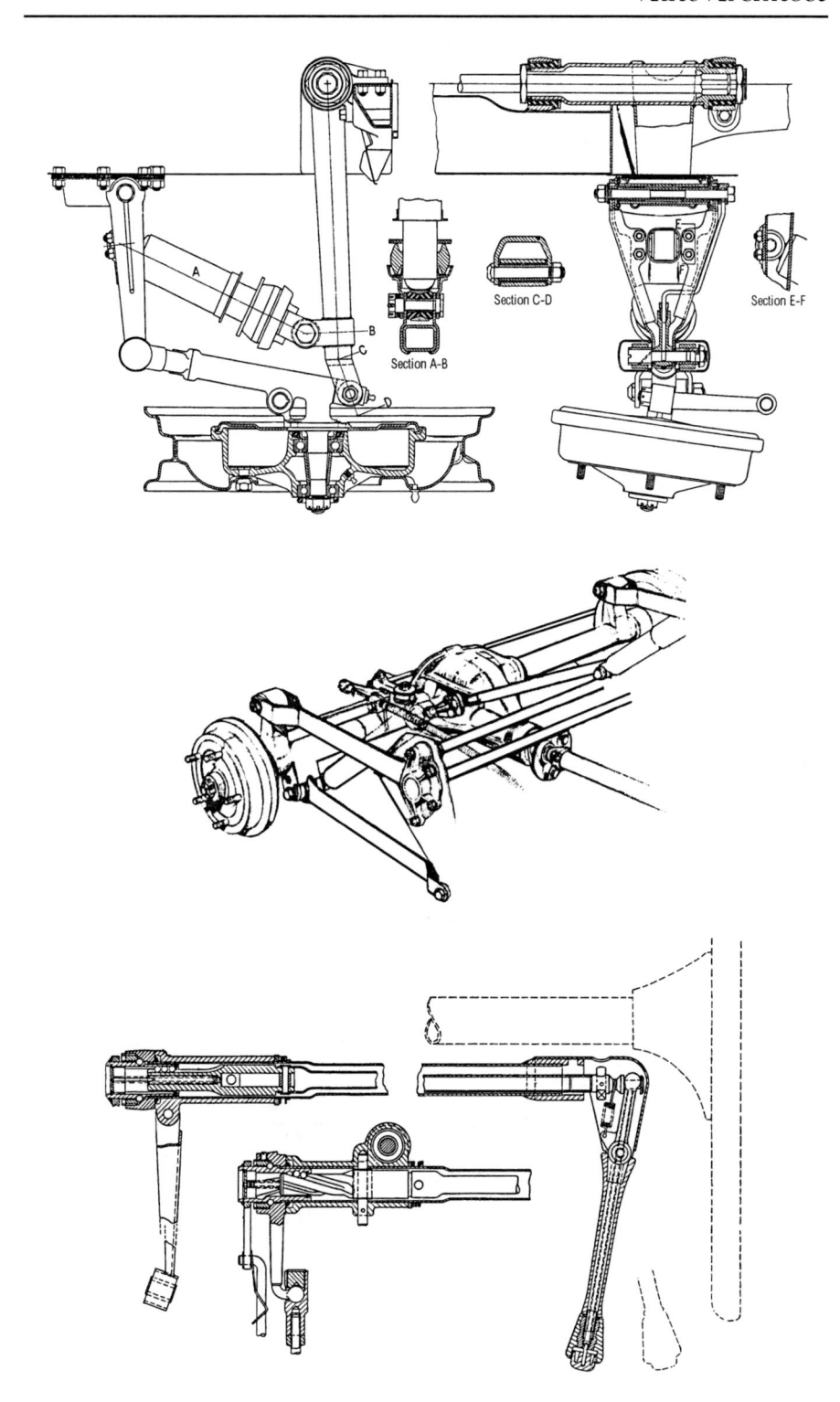

Section C-D

Section A-B

Section E-F

Auto-Architect

too, with the VW; a strong box with compact engine and absorbent suspension. (The promising Fedden prototype came to grief during strenuous testing in 1946.) I made preliminary rough drawings with stubby, attenuated front wings finishing before the leading edge of the door. In chasing minimum tooling costs I realised that front left and right rear door panels (and vice versa) could be made interchangeable, with fabricated window frames added separately. Enclosed rear wheels were a streamline touch. We did not have modelling clays so Jowett's one skilled craftsman in wood fashioned a 1/4-scale model from a solid block. No wind tunnel was available but, having pushed the rear seat position forward, the natural curve for the rear of the body assumed a reasonable shape. This model and the drawings were shown to the directors, who approved them, and I was given the go-ahead to build a running prototype.

A larger office was provided and the services of two young draughtsmen, Howarth and Watkins, to work on the engine and mechanical units. At a later stage, when body details were necessary, Reg. Korner joined the group. He had come to Jowett in 1936, having started working with traditional London coachbuilders and then moved to mass production in the Midlands. He had modernised the pre-war bodies and was later responsible for designing the curvaceous Jupiter bodywork.

The type of construction was a major problem as the company had neither the facilities, nor apparently the cash, to go for an all-steel body. Some time was spent in investigating various alternatives, such as plastics, laminated wood, and aluminium that would be pressed using rubber dies as in the aircraft

The only Fedden sleeve-valve radial-engined car built, photographed in Cheltenham in 1945/46. The car was destroyed in an accident; it somersaulted during high-speed testing.

industry, but none of these was viable. In a metal body, the roof, front and rear bulkheads, and door aperture frames contribute to the torsional stiffness of the whole structure. As I could not rely on these I decided to have a base frame consisting of two massive, but straight, light box-section members united by an equally massive cross-member, and by the welded-in front and rear bulkheads. This more than compensated for the possible deficiency in torsional stiffness. An additional advantage was that it gave very positive location of the front and rear road wheels, to which I partly ascribed the good road handling of the car. The front suspension lower arms were directly pivoted on this base frame, as were the radius arms for the rear axle. It also supported the floor of the passenger compartment, which was a plywood sheet in order to save the cost of a steel floor. A full-size wooden mock-up was built, and the external shape approved. The front wings were lengthened to flow into the doors and helmet-like rear wings enclosed the top of the wheels in place of the full spats. A seating buck was also built to check details of the seats, gear lever, steering wheel and, particularly, the position of the pedals.

One way of retaining or increasing the passenger space in a car, which was smaller than earlier designs, was to bring the front wheels back towards the driver. This meant that the front wheel arches intruded into the passenger compartment and the pedals had to be placed nearer the centre. This was a trend

An early full-sized mock-up of the Javelin showing the 'V'-shaped windscreen and thinner door pillars. By this time the original proposal of having short front wings and interchangeable door panels had been discarded.

Auto-Architect

already set by Volkswagen and since widely followed. With right-hand drive, the driver needs to sit with his legs at an angle, and those with long legs may have to slightly twist the knee to reach the accelerator pedal. It is less of a problem with left-hand drive where the right leg is extended to reach the pedal. I asked all the directors to sit in the buck. I wanted to be sure from the start that they understood, and were happy about, this small deficiency which came with the advantageous decrease in overall length. To expedite building the first running prototype, and before the question of the type of body construction was settled (which in fact was finally settled by the promise of this prototype), I used an aluminium roof panel clinched to the cant rail; this did give a satisfactory torsional stiffness figure for the whole body but would not have been practical on the production volume in view. Aluminium was also used for the bonnet, door outer panels, and boot lid outer panels. Of course, these did not affect the stiffness figure.

Because the R.A.C. Horsepower Tax favouring a small cylinder bore was in force at the time, the new engine was originally designed as a 1.2-litre unit for the home market, with a larger bore version of 1.5-litre capacity for export. More detailed estimates of performance then showed that the smaller engine would not give the target aimed for, so the 1.2-litre unit was not proceeded with after the first prototype. Too late in the development stage the Government changed the taxation rating to a capacity basis. Had I known this I would have chosen a

The second Javelin prototype photographed with a fashionably-dressed model in London. I notice that I have dated the picture 18th March 1945 and written that 'This is the new car – I hope you like it. It seats 5 in comfort and holds 5 good-sized suitcases. It has only a 10hp engine but will do 75mph, 32mpg on petrol, and is quite lively. It's very comfortable to ride in and holds the road marvellously. It has many features in common with the 'Deroy' and experience in that has helped me greatly with this. Plans are going ahead to make this in thousands!'

square or over-square bore ratio (where the bore width was the same as, or greater than, the piston stroke). This would have reduced the engine width and eased crankshaft design.

The first engine had a single-piece, cast-iron cylinder block with a two-bearing crankshaft. I should have known better. It was very noisy and it was obvious that a three-bearing crankshaft was needed. With hindsight I should have kept the cast-iron block and used a barrel type centre main bearing. The engine was redesigned as a three-bearing unit with a single-piece, aluminium block, but again, due to crankcase flexing, it did not cure the noise problem. I decided to abandon the one-piece cylinder block/crankcase and redesign it as an aluminium two-piece, split on the vertical longitudinal centre line and bolted together. This, while being slightly more expensive, largely solved the noise problem. It did, however, open up the possibility of gravity die casting the two halves. It so happened that Renfrew Foundries, part of the Alcan Group who produced Rolls-Royce Merlin engine crankcases, were looking for post-war products suitable for their Glasgow foundry. They heard of the Jowett project and through their technical representative co-operated with us in the design of the die-cast cylinder block. I believe the Javelin was the first UK production car to have such a unit.

The two cylinder heads were identical and these were in cast iron to avoid using valve seating inserts that would have been needed had they been of aluminium. Combustion chambers were of conventional wedge shape, and the sparking plugs were on the top face, necessitating a carefully-designed plastic cover to keep out water. Due to the width of the engine, and because the single camshaft was on the crankshaft centre line, the push rods were longer than on a four-in-line engine. Partly for this reason I used zero-lash hydraulic tappets which were just becoming available in this country, although these were costly. However, their service history was not good and they had to be replaced by a solid type. Because of the length and weight of the pushrods and valve gear this engine was not designed for high speeds; for this reason I was always apprehensive about it when used in a sports car.

While this engine development was going on, the first prototype, completed and running in mid 1944, was being used to test and develop the suspension, brakes, controls and other components. It was the suspension medium which gave most cause for concern. Mindful of the problems with the universal joints on the independent rear suspension of the Deroy, I had chosen a rigid hypoid rear axle. Also based on Deroy experience, I had elected to use torsion bars with the independent front suspension and identical bars at the rear. These could be neatly housed without encroaching too much on the passenger and luggage space. The first torsion bars were enlarged at each end and splined, according to the practice then current. On initial road tests they failed at a very low

Auto-Architect

mileage, as they did at very few cycles on a rig fatigue test. The fractures always occurred where the machine splines ran out into the plain bar, i.e. where there was a stress raiser. To eliminate this was difficult as I could not increase the diameter of the splined end since it would alter many adjacent parts. Several different steel and heat treatments were tried to no avail. It was not until I hit on the idea of using octagonal ends in place of splines (hence no stress raisers) that the problem was solved. I don't think another torsion bar ever broke on test or in service. It is interesting that Chrysler, who were developing a torsion bar suspension for their US models, adopted the hexagon bar after examining the Javelin which they were the first to import into the USA.

A full programme of road testing continued on a tough circuit in the Yorkshire Dales using numbers two and three prototype cars. I played my part with the daily ten mile drive to bracing Ilkley where we had moved in 1945 (after finishing my Home Guard duties as Corporal Palmer!). I was greatly assisted by Horace Grimley, a nephew of William Jowett, who had started with the Company in 1921 and become a capable and discerning test and development engineer. Horace was a tower of strength and with his team, Illingworth and Metcalfe,

The second prototype again, this time in my hands on Sutton Bank – a well-known venue for speed hill-climbing in the 1920s. In this picture the lower radiator grille, narrow window frames, and divided windscreen are evident.

virtually created the prototype from raw materials. Time and again they worked through the evenings and at weekends. Compared with modern practice our efforts were still inadequate, but faults were found and corrected. An example was a broken front suspension lower wishbone which led to the redesign of a stronger component. The success of the project was in no small measure due to their contribution. They worked wonders on a small budget and with meagre facilities.

I had kept the Deroy. With the saloon at the testing stage there was time to replace the low output Scammell unit with a Javelin engine. I had been staggered by the artistry of the Italian body builders when I visited the Turin Show and with the help of a Jowett panel beater, we managed to replace the external panels of the Deroy with ones of a full-width streamlined shape.

(Left, below and page 48) The Deroy in its second form. The Scammell engine was replaced by a Javelin unit that gave the car far better performance. I was keen at the time on all-enveloping bodies and, with the help of a Jowett panel beater, fitted the Deroy with a body that looked very modern at the time. It is a pity that the car does not seem to have survived.

Auto-Architect

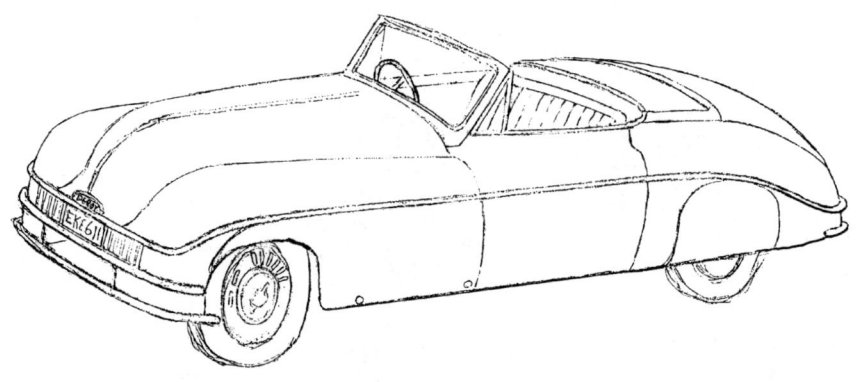

Chapter Six

Producing the Javelin

Whilst work on the Javelin proceeded, there were changes to the financial structure of the Company. After the holding of the Jowett brothers had been purchased by Charles Clore (known as Santa Clore to the rank and file!) he in turn sold out to merchant bankers Lazard in May 1947. The Board changes in June brought in Wilfred Sainsbury as Lazard's nominee and financial director, and George Wansbrough as Chairman. Peter Reilly and Harry Woodhead remained as joint managing directors until Peter's forced departure in January 1949. The other directors blamed him for delay in the car programme and may have preferred him to have been full-time at the factory.

It is always possible to work still longer hours but, with the facilities and materials then available, I know it would have been difficult to further advance the project. It's true that Peter had taken his large family back to their house at Teddington. It had been taken over by Shell whilst the Reillys rented Bramhope Manor near the factory during the war years. But it wasn't just the attraction of a house they liked by the river. Peter felt that he needed to be based in London for his battles to convince the Ministry that Jowett's new, if as yet unproven, car merited the steel supplies and the support which were more readily given to older designs like the Nuffield products. In those early post-war years, the system for steel allocation depended on past production; in the case of Jowett this was minimal compared with the Javelin's potential. Peter was also developing export and other contacts and therefore based himself at the offices above the Albemarle Street showroom. However, he came up to the factory each week and we still maintained a close partnership.

I got on well with George Wansbrough, a man of many talents. An Eton scholar and Captain of the School, he had subsequently stroked the Cambridge Eight in the 1925 Boat Race. He had then worked for the Benson merchant bank before joining manufacturing industry at director level in 1934. That same year he had competed in the RAC Rally, driving a Siddeley Special with a futuristic streamline saloon body built by Lancefield. By 1947 he was chairman of Reyrolle and on the board of British Power and Light, Morphy Richards, and C.A. Parsons.

Auto-Architect

He was well known in the City as a director of the Bank of England and of Mercantile Credit. He was also sufficiently respected by the Labour Government to have been appointed to the Advisory Council for the Motor Industry, with the likes of Miles Thomas (Nuffield), Charles Bartlett (Vauxhall), Spencer Wilks (Rover), Leonard Lord, and Reginald Rootes. George was able to organise a visit to Idle by the young President of the Board of Trade. I remember joining George, Harry Woodhead, and Harold Wilson for lunch at the Midland Hotel in Bradford. The Minister toured the factory and we did subsequently receive adequate supplies of steel.

The introduction of Lazards had also opened the door for new capital. As it happened Briggs Motor Bodies had a plant at Doncaster producing pressings for various war purposes and were coming to the end of their contracts. They were looking for new work so what could be more obvious than for Jowett to source the complete Javelin body, painted and trimmed, from Briggs – provided the price was right and that Lazards provided the capital. It was so, and I attended the final meeting down at Dagenham with Peter at which the contract was signed,

Charles Calcott (Peter) Reilly, Managing Director, with early Javelin outside the Jowett London showrooms in Albemarle Street in 1947.

and where I met the Briggs technical staff with whom I had to work. The general design of the car was altered very little and then only by the adoption of pressed steel construction techniques in place of semi hand-made methods. This mainly affected door construction, resulting in thicker sections for the window frames which reduced visibility. The underbody, or frame, with the plywood floor and all internal panels were changed only in minor details but external panels, where we

> ## 'PETER' REILLY'S DISMISSAL
>
> Speaking in 1997 Pam Chase, Peter Reilly's daughter, clearly remembers her father's dismissal in January 1949. He had organised the Jowett factory for war production, he had initiated the Javelin project, he'd sought out Gerald, and the Javelin was an integral part of all their lives. Pam and her sister had both learned to drive and passed their tests on one of the early cars. When they were all working in the London office one of them acted as chauffeur for the drive from Teddington to Albemarle Street. Pam recalled the easy and forgiving nature of the design. One icy morning she misjudged the conditions and the car slid into a kerb but there was no ensuing drama. Her father just continued his work in the comfortable back seat.
>
> The family couldn't believe what had happened when Peter returned from that fateful Board Meeting when the voting went against him. Her father was bowled over but didn't make much comment, he just got on with the next stage of making a living (with the Cyclemaster, a power assisted cycle with the engine in the back wheel). However, their forceful step-mother (their mother had died years before) made her feelings quite clear and gave George Wansbrough a piece of her mind in no uncertain manner. In fact, George could have done nothing with the other Directors united against Peter.
>
> Pam continued for a while with her work on the export side, obtaining permits and allocating vehicles to available shipping space. She remembers the remarks on visits to the factory after the dismissal: 'There is no one on the Board now except the financiers' and 'The money-bags have a one-track mind'. Pam has two other pertinent memories, of the way the cars were raced to the docks by their delivery drivers (which could not have helped the engine problems), and of the brusque way in which one of the Yorkshire directors ordered her to have the tyres (then like gold dust) taken off a line of Bradford vans which already had designated shipping space.

had used aluminium on the prototypes, were changed to steel, with a consequent increase in weight of about 150lbs. This resulted in a shipping weight of 2,150lbs, still a reasonable figure for a six-passenger car.

1947 and 1948 were years of frantic hard work in finalising all details, correcting any failures, and preparing drawings and schedules for production release. Liaison with Briggs at Dagenham each fortnight was essential to check progress on the production prototype for the Doncaster plant. We had to make sure no unacceptable changes were made and to impress on them to keep to our weight target. We always completed the 400 mile round trip there and back in one day, at first using a Citroen Light Fifteen owned by the factory and then one of the prototype Javelins. There was little traffic on the old A1 from Doncaster to Barnett. During those years of reduced fuel imports (to save payments to other countries) we had our essential user petrol coupons whilst private owners were restricted to the parsimonious basic ration. This was set at nine gallons a month for a small car in January 1947, removed altogether the following winter, and then reintroduced at three gallons a month in June 1948. At this time we were approached by the Triplex Glass Company to adopt a curved windscreen in place of a divided vee-screen. They had installed new plant to produce them which was to come on stream co-incident with the launch of the Javelin. We naturally accepted their offer to have such an up-to-date feature, so the Javelin became the first UK production car to be so fitted.

In April 1947 the management team had been strengthened by the appointment of Frank Salter as Production Director. He was ex-General Motors Antwerp and for some years had been Sir John Black's right-hand man at Standard. His innovative approach to production problems is well illustrated by his roll-over feature of the Javelin assembly track, whereby mechanical units were assembled with the body upside down. He contributed many other useful suggestions to ease the production of components and always did so in a diplomatic way. Charles Grandfield was another recruit in 1948. An Austin apprentice in 1931, he then served in the army from 1939-45 and become the REME Colonel in charge of mechanical transport for Field Marshal Montgomery's 21st Army Group. After two years with Roy Fedden he became Jowett's engineering manager with full responsibility for getting production releases finalised and out on time.

The Bradford continued to bridge the gap between the end of war contracts and the start of passenger car production. It had a simple wood-framed, steel-panelled body and in the 5-8cwt class found a ready sale in a transport starved country. I had no part in the project as there was practically no design work involved. The complete bodies, painted and trimmed, were supplied by Briggs from their Doncaster plant. Over 38,000 of these were produced before it was phased out in 1953 and they provided a valuable source of income in a financially difficult time for the company. John Baldwin had joined as publicity manager in 1946. He was based in the London showroom and glad to have something to sell until the Javelin was available. Bradfords were exported to Australia, New Zealand, South Africa and South America, as well as to many European countries.

It was now time to introduce the public to my creation. To celebrate the Motor Industry's Fifty Year Jubilee, the Society of Motor Manufacturers and Traders organised a display in Regent's Park on Saturday 27th July 1946. This was followed by a cavalcade - Park Lane, Piccadilly, Shaftesbury Avenue, Oxford Street, and back to the Park. We entered EAK 771, the prototype with the Triplex screen, and the *Autocar* commented 'Spectators were puzzled and then surprised by the discovery that this was the 1946 Jowett'. The factory also entered a 1938 10hp saloon and the 1913 6hp light car, with John Baldwin manfully coping with working out which way to turn the lever before each change of direction on the tiller-steered machine.

After some preview leaflets and a few leaked press releases during the winter, the Javelin was fully described in both the weekly magazines in May 1947. I am diffident about including press reports but, for the record, *The Autocar* did comment after a 700 mile trial in the hand-built EKW 303 still fitted with a vee-screen: 'The suspension is of exceptional merit giving a smooth ride without any

Javelin and Bradford outside the Albemarle Street London Showroom.

Auto-Architect

(Above) George Wansbrough, who was appointed as the Chairman of Jowett in 1947, with FAK 698 at Hyde Gate, Long Sutton before his departure with the car to America. Peter Reilly and George travelled across the Atlantic in the Queen Elizabeth, and then drove the car to Detroit and Windsor, Ontario.

(Below) FKU 372, photographed in Sweden with the transparent roof panel that was too costly to be fitted to production cars.

(Right) Javelin chassis number 5209, GAK 335, was one of the 1949 Motor Show cars. As an early example it has the smaller, 5.5in headlights, one-piece chromed radiator grille, and the badge on the top of the bonnet.

tendency to pitching. The Javelin can be summed up as possessing an extremely lively performance for a car of 1.5-litres'. *The Motor* reported 'a combination of qualities rendering the car unrivalled in its own field'. As there was no 1947 London Show, the international debut was delayed until the Brussels Salon in February 1948. FAK 573 was driven to the Geneva Show in March and was then the subject of the full *Autocar* road test. At the end of April, George and Peter escorted FAK 698 across the Atlantic in the Queen Elizabeth and drove to Detroit and Windsor, Ontario. We had achieved a product which attracted what we would now call rave notices.

Stands were taken at European shows and the year concluded with the first British showing, a Golden Sand and a Turquoise saloon ('a design of which British industry can be proud') and the experimental roadster, at Earls Court in October. The attempt at a roadster was nothing to do with the factory. It was a private venture by Peter helped by one of the London agents. Harking back to his Gordon England days, he was thinking of the same market as the designers of the open three-seater Riley, but the altered body didn't have sufficient structural strength and the car did not go into production. I felt the proportions were wrong, the long tail out of place on a concept with a short bonnet.

I was sorry that the tooling costs for the transparent roof panel fitted to FKU 372 (with which Peter toured Sweden in September) proved too high. Laurence Pomeroy of *The Motor* also drove this Javelin and wrote in his *Account*

Auto-Architect

Rendered: 'most favourably impressed by the Perspex roof. This really does combine many of the virtues of open and closed motoring and is an extra I can most warmly commend to those who retain a lingering affection for real visibility'. Pomeroy also commented that the car could be best described as being an anglicised and, in many respects, improved Lancia Aprilia.

For our summer holidays in 1948 and 1949 we had driven our own early production duotone beige and brown car FKU 279, which was christened Chocolate by Celia, to St Tropez. Uncrowded and not yet fashionable after the war years, we found furnished rooms directly overlooking the harbour. One year we looked across to Erroll Flynn's yacht moored opposite. Expeditions were to the Plage de la Briande where we lunched in the shade of the pines, looking out over deserted sands to the sea. In 1948 we were six up (though two were children)

plus luggage and the Javelin remained comfortable over imperfect surfaces and trouble free. The next year we returned to discuss overheating problems with Joseph Stierli, who had the Swiss concession.

The business currency allowance added to the holiday allowance was stretched to cover two days in a cheap hotel. We long remembered, after the food shortages in England, the sumptuous dinner with the Stierlis in a village restaurant by the lake. Unfortunately, and also long remembered, Madame Stierli was taken ill halfway through this wonderful meal.

The chocolate car continued to run sweetly and gave no cause for alarm when we tackled the Col d'Iseran. However, owners who used the Forclaz between Martigny and Chamonix, then the steepest pass in Europe, were reporting problems for which radiator modifications were put in hand on our return. Hubert Patthey at Neuchatel added the Javelin to his AC, Alvis, Bristol and Nuffield agencies. In 1947 he had driven the first production car, FAK 111, back to Yorkshire and returned with the first of many Bradford vans sold in Switzerland. In Brussels we were represented by the formidable Madame Sterckx.

(Left and above) At the Bradford factory an ingenious production line was built that allowed Javelin under-body components to be easily installed.

Auto-Architect

(Above) The Bradford Jowett factory in 1951. The cables are to power the trolley buses that passed the front of the building, convenient for the workers few of whom had cars themselves at this time.

(Below) JAK 75 was the 1952 Motor Show car. Note the 7in headlights and the 'Jowett Javelin' badge above the two-part radiator grille.

(Above) The duotone early production Javelin, chassis D8 PA 179, we drove to St. Tropez on holiday. The yacht on the right belonged to Errol Flynn, the Hollywood film star!

Auto-Architect

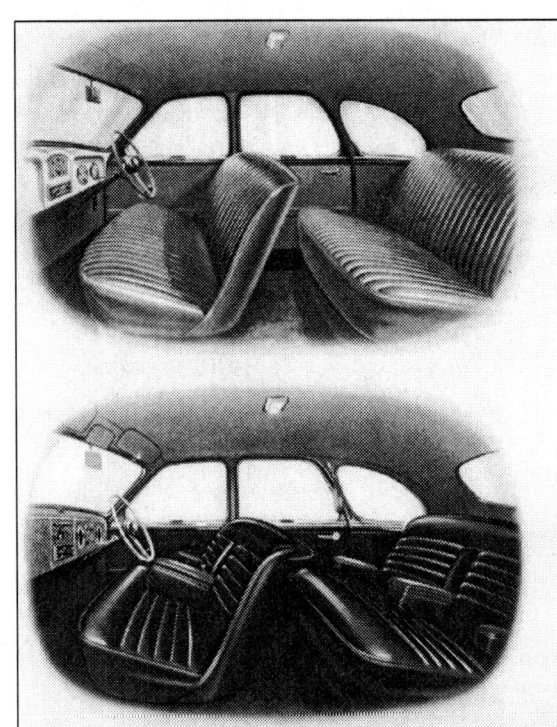

interior of JAVELIN *saloon*

You will approve of the well-planned layout of this roomy car, which is remarkable for the amount of space given to both front and *rear* seat passengers. Plenty of leg room, a flat floor front and rear, and real inter-axle seating are main features. The base of the rear seat squab is actually 15 inches in front of the rear axle. Essential instruments are grouped in front of the driver, and both driver and passengers notice the remarkable all-round vision. The interior upholstery is in P.V.C. — good-looking and extremely hard-wearing.

interior of JAVELIN *saloon de luxe*

Here is an interior of outstanding finish and comfort. Folding arm-rests for the driver and all passengers, a front squab which tilts as the seat slides forward, and real spaciousness. The full instrument panel of walnut grain and the hide upholstery have a conservative appeal to the discriminating owner. Arm-slings are provided for the rear passengers, and the front passenger can use a make-up mirror on the sun visor. The rear side arm-rests are detachable for making use of the full seat width when extra elbow-room is wanted. To add to the pleasures of the journey, car heater, map pocket and detachable picnic tray are included. Under the dashboard is fitted a spacious draw-type ashtray. H.M.V. Radiomobile car radio extra.

Chapter Seven

Monte Carlo and the Return to Oxford

In November 1948 it was announced that the famous Monte Carlo Rally would definitely be held in January for the first time *post bellum*. We at the works had never viewed the Javelin as a rally car as it had been designed purely as a utility vehicle. It was a well known Yorkshire motor sportsman, Tommy Wise, who saw its potential and he persuaded the management to sell him the latest prototype, which he entered into the Rally with T.C. (Cuth) Harrison as co-driver. He very kindly asked me to join them as second co-driver. 'It would be a salutary experience for you as the designer', he said. I jumped at the opportunity and the experience has remained a vivid memory with me ever since then.

The British section of the Rally started at Glasgow, the authorities having provided all competitors with petrol coupons to drive from home to the start and down to the South. Our first stop was at Tommy's house at Guiseley, where our wives fuelled us up for the night section to Folkestone. Over the Channel the 'wicked' French had laid on a champagne party at Calais, perhaps to delay the naive British entry? Suddenly Tommy yelled, 'We're an hour behind schedule'. I shall never forget that journey across the atrocious roads of Northern France to Luxembourg. They wouldn't let me drive as they said I'd treat the car too gently. So it was bang, bang, bang from pothole to pothole, me in agony wondering what would break first on the car. But we made it to Luxembourg with minutes to spare – and with the car in one piece. The route then led north to Holland (Nijmegen, Amsterdam, and The Hague) through Antwerp and Brussels to Paris, where we arrived about 9.00pm on the third night with the control under the Eiffel Tower.

The enthusiasm of the French populace was extraordinary and my recollection is of cheering crowds lining the streets of every sizeable town, waving and urging you on. It was an all night stretch through France until the dawn came up at Lyon; and with it the dreaded fog. Tommy drove through it but we were obviously losing time, so on turning east towards the mountains and clear weather we had to make it up. I took over and on the long straights down to Sisteron and Digne it was flat out. 'Faster', yelled Tommy from the back seat. 'It won't b..... well go any faster', I yelled back to loud laughter from the two of

Auto-Architect

them; me wondering if the engine would survive. Cuth took over at Digne and we came up to the first pass, the Col des Leques, which was not high but difficult with lots of hairpin bends. Cuth yelled, 'Shall I let her rip Tommy?' 'Yes', came the reply, and I immediately groped for the door exit handle as we approached the first bend at a madly high speed. Cuth literally threw the car round that bend in four wheel drift, me having kittens wondering what would break. I suddenly said to myself, 'this chap knows what he is doing', and was completely calm as he repeated the performance at every bend. I timed him over the pass and he averaged 40mph! Phew! I had never experienced this kind of driving before and was never surprised that Cuth became one of our top Grand Prix and rally drivers in later years.

Really comfortable 3-abreast seating; all controls and dials opposite the driver, clear floor space and good visibility opposite the passengers. The saloon de luxe has a central folding armrest in both the front and the rear seats, also elbow rests.

Real inter-axle seating, for the short engine permits a forward body position, flat rear floor and plenty of leg room. The rear passengers, as well as the front, have a bounce-free ride. And the roomy boot is not all overhang.

Ground clearance designed for rutted tracks, bad pot holes and the worst surfaces that cars can travel. A minimum of 7¾ in. is just one of the many essential overseas features incorporated in the 'world-market' JAVELIN.

Out of sight, out of mind, but clean, space saving and thief-proof. The spare wheel is carried on a tray below the tail. Turn a nut under the luggage locker lid and have your spare wheel out in seconds.

And so into Castellane and down to the coast at Nice and on to Monte Carlo. The car and I had survived and we'd lost no marks on the road section. A few hours of rest and relaxation followed and then came the eliminating test, which was a speed hill climb up the Mont des Mules. Tommy made one or two practice runs and was satisfied that he'd be well placed. He did his two climbs, made good time and was confident. To our astonishment when the results were posted up, Gatsonides, the Dutchman in a Hillman Minx, was placed first in the 1.5-litre class and we were third, and 27th overall in the Rally. I said 'I don't believe it; let's go and look at the detail figures', but there was not time before the prize giving, so Gatsonides was presented with the Riviera Cup and £75 by Prince Rainier and we got a miserable £35. We then rushed to examine the detail times and I found a simple arithmetic error had been made in our result. A protest was immediately lodged which had to be upheld by the rather confused and red-faced officials. 'Ah monsieurs', they said, 'this is the first rally since the war and we are out of practice'. The revised result put us way ahead at the top of the 1.5-litre class and 14th overall. The Riviera Cup was grabbed back from the hapless Gatsonides and given to Tommy, but he'd quickly disposed of his £75, so they had to make it up to us. Another Javelin, driven by Ron Smith and Horace Grimley, was 22nd and third in class. Three further examples (Turner, Hume and Miller) were amongst the later finishers.

As a postscript to the Rally, I had hoped to have a day or two basking in the sun. Not a bit of it! We left that evening and spent the night at Aix. Setting off at the crack of dawn, we eventually bedded down 500 miles later, near midnight,

features of the $1\frac{1}{2}$ *litre* JOWETT JAVELIN

Correctly adjustable driving seat (JAVELIN saloon de luxe). Turn an easily accessible handle below the seat and you raise and incline the squab; wind the seat forward or back to give safety and comfort for drivers of all heights, men and women.

Outstanding engine accessibility that the owner-driver has demanded for years: open the bonnet and the top half of the grille pivots up; the lower half of the grille is easily removable for major adjustments.

at Amiens. I well remember that all the hotels were closed. We saw a light in a basement shop; Tommy just walked in to find an astonished baker at work and he managed to persuade the chap to give us beds either in his own or a neighbour's house. Those who remember the state of the French roads in 1949 will agree that that was a good day's motoring. The car came through intact which was a great relief to me. It was a real privilege to have driven with Tommy Wise and Cuth Harrison – they were two fine Yorkshiremen.

No company could have wished for greater publicity at the debut of a new model than this win at Monte Carlo. It was followed in July by an equally spectacular win at Spa in the Belgian 24-hour race. Anthony Hume, who had also driven a Javelin on the Rally, had then read the Export Report by Belgian journalist, Paul Frère, in the 23rd February issue of *The Motor*. Frère noted that the Javelin stand at the Brussels show had not attracted much attention, concluded that the car was too little known and suggested running a team of three cars in the touring class at Spa. This led Anthony to ask us to enter a team. We decided that there was not enough time to prepare three machines but that we would lend one production car to be driven by Anthony, together with Tommy Wisdom.

This was prepared in accordance with our service bulletin recommendations for competition, which involved stripping the engine and careful reassembly, raised compression, larger carburettors, higher gear ratios, racing tyres, and stiffer shock absorbers. The car ran like clockwork, needing no repairs or adjustments. It won the 2-litre Touring Class covering over 1,500 miles at 65.5mph, and covered a greater distance than all the 1.5 cars (both sports and touring), and the much larger 4-litre touring cars. Other rally successes followed, so that the car was beginning to be known by its semi-sporting rather than its utilitarian qualities, which were the original design objectives. This was disturbing to me since it might lead to its getting a reputation for unreliability. I knew there was still a lot of development work to be done. (Apart from the use of Javelin mechanical units, including the torsion bar suspension, I had nothing to do with the Jowett Jupiter. Von Eberhorst's steel tube frame was an idealistic layout for a limited production sporting car but it was not easy to manufacture.)

With the successful launch of the Javelin, the very favourable press reception it received, and the fact that my name was closely connected with it as the designer, I found myself the recipient of a certain amount of acclaim in the industry. I was getting known! Therefore, it should not have surprised me when it was whispered in my ear by a mutually friendly trade representative that my former company in Oxford might be interested in re-employing me in a design capacity for some of their future projects. My immediate reaction was one of surprise and elation. I had fulfilled my contract with Jowett and had produced for them a worthy successor to their pre-war range of cars, which with vigorous

Auto-Architect

(Above) 1949 Monte Carlo Rally. With Cuthbert Harrison, Tommy Wise, and the Javelin we drove in this event. We were placed first in class after an arithmetical error in the results had been cleared up.

(Below) A car was entered in the 24hr race at Spa in Belgium where it won the two-litre touring class, driven by Anthony Hume and Tommy Wisdom.

Auto-Architect

development and competently produced, should have a sales life of several years before major changes were necessary. As I could see no future design project in view at Jowett, and I was a designer not a development engineer, I felt justified in pursuing this approach from Oxford and fixed up a meeting with the Nuffield Deputy Chairman. The concept team had already been dispersed by the Lazard people, with Peter Reilly dismissed and the ever helpful Steven Poole pushed into taking retirement.

Unfortunately, after I had left, the development and production of the Javelin did not continue as I had hoped. The gearbox was a salutary example. There were no serious problems with the boxes built by Meadows. Then, to save money, Jowett decided to make their own which jumped out of gear. There was insufficient pressure to keep the sliding dogs engaged when in top gear, which I think was a consequence of the worn state of Jowett's tooling. At Vauxhall there had been a similar problem with gear shapers cutting very slightly tapered dog teeth. These tapered teeth caused an axial load thus forcing the gears out of engagement.

The expensive to produce, steel tube chassis of the Jupiter was the work of Von Eberhorst, I had no part in the design other than of the main Javelin mechanical units and torsion bar suspension. This car was built for the 1951 Le Mans race.

Chapter Eight

Nuffield Again: Magnettes and Others

The machinations for the English motor industry in the 1920s and 30s are too well known and chronicled for me to add anything here, particularly as a comparative outsider at that time. Two of the early companies, Wolseley and Riley, were in financial difficulties. To prevent them being acquired by his arch rival Herbert Austin, they had been purchased by William Morris who was then faced with the problem of amalgamating them into his ever growing empire. His strategy was to appoint a deputy chairman to relieve him of second line responsibility on organisation and general product and production decisions. In this he was only partly successful. After one or two in-house appointees he recruited L.P. Lord, an Austin executive, to the position. It was not a success, due to the incompatible temperaments of the two men, and L.P. returned to Austin in 1936. He was replaced by Oliver Boden who sadly died in 1940.

The next and most successful deputy was Sir Miles Thomas, who held the position through the war years and lost it in 1947, together with several other directors. At the time, it was generally thought that this boardroom purge was a preliminary to the amalgamation of the Austin and Morris companies, which eventually took place in 1951 when the British Motor Corporation was established. In the period from the departure of Miles Thomas to the formation of the BMC, Morris (or Lord Nuffield as he had become) had appointed R.F. Hanks as his deputy to sort out the amalgamation of his empire, and it was to him that I reported for an interview. A.V. Oak, my previous boss, had remained as technical director but was near retirement. Alec Issigonis had produced the small four-passenger Minor which was starting its astonishingly successful production run, but they had no one responsible for the design of the future M.G., Riley and Wolseley models - would I like the job? This was a glittering prize to be offered and I had no hesitation in accepting it; realising at the same time the responsibility and hard work involved.

On my return to Oxford in July 1949, I was allocated a small drawing office and staff in the Cowley complex where all the layout and styling work was done. I shared a room with Harry Rush, designer of the Riley RM range. It was a little awkward sharing with a man whose job you were taking over, but Rush spent over half of each week in Coventry and was tragically killed in a road accident in December, so I really saw little of him. Alec Issigonis had his office next door. Many people regard the Morris Minor as his *magnum opus*. Certainly it was from a commercial point of view for it earned Morris Motors a profit of about £40 for every vehicle leaving the factory gates. In contrast the later and better known Mini could not have earned anything like this figure, if it was profitable at all. Indeed, in the opinion of many people, it led to the insolvency of the company. I well remember some years later at a presentation of new Vauxhall models to General Motors top brass at which the newly announced Mini was on show. Scant attention was shown it and the remark was made that 'You can take value out of a product quicker than you can reduce its price'. In other words, it is easier and quicker to adjust profit by altering specifications and equipment than to push price changes through the company's marketing and sales organisations.

I got on well with Alec Issigonis – we shared an arrogance common to designers, he possessing it more deservingly and openly than me, and it increased with his success. Some years later there was an article about the Minor in one of the weeklies. It reminded me of a Coward play; written by Noel, acted by Noel, directed by Noel. I'm not denigrating Issigonis. I admired him immensely but Vic Oak was not mentioned at all and he was the responsible director. Vic had always been a good friend and I'm glad that my letter recording his contribution was published. Issigonis was already obsessed with the concept of a very small car, which he called his 'charwoman's car', and often talked to me about it.

Above and right) Wolseley 4/44. Side view and body construction. When designing the body shell I made use of the propshaft tunnel and sills to transmit the loads from the front suspension to the main body structure. To give it sufficient strength to carry these loads, the propshaft tunnel was deeper than would otherwise have been necessary.

Auto-Architect

My brief at Cowley was to design new saloon cars for the M.G. and Riley ranges, and two saloon models for the Wolseley range - one a four-seater and one a six-seater. All of these were to use standard BMC mechanical components, such as engines, transmissions, and rear axles. I had a completely free hand so far as the size and styling were concerned. I had to assess in my own mind the type of prospective purchaser to which each should appeal, their social standing and lifestyle, as well as maintaining the character of the car which had been established by its provenance. The situation was further complicated by the uncertainty of the engine programme. Early talks were apparently taking place which were preliminary to the merger of Austin and Morris, the outcome of which was that the Morris 1¼-litre engine was to be phased out and replaced by a new 1½-litre engine (Austin B-type) and a new six-cylinder, 3-litre engine was under development at Austin (the C-type) to replace both the Morris six-cylinder and Riley 2½-litre engines.

Early in the programme it became clear to me that, both from a financial and a timing point of view, four complete new body shells were not viable for the modest production volumes planned. Therefore, I proposed that there should be two basic body shells, one for the four-passenger models and one for the six-passenger models; the former for the M.G. and small Wolseley, the latter for the Riley and large Wolseley. Marked difference in appearance between the three makes could be achieved in several ways.

Auto-Architect

For instance, by altering the free camber of the road springs the standing height of the two cars could be different. The M.G., in my proposal, was two inches lower than the Wolseley, giving it a low slung, more sporting appearance. This meant a different outer sill panel and different cut-outs in the wing panels for adequate tyre clearance. At the front end changes were made to bumpers, lights, panels and radiator grilles, each of the latter being of the characteristic shape of the marque which provided strong, unmistakable identity. The interior treatment provided greater scope for changes; the M.G. having a floor mounted gear lever between the front seats, together with the adjacent hand brake, while the 4/44 had a steering column mounted gear-lever and under-dash handbrake, freeing the space between the two front seats for accommodating a small child's seat, should this be desired.

The treatment of the two instrument panels proposed was completely different. In the case of the M.G. the speedometer and subsidiary gauges were placed inside a semi-octagonal housing in front of the driver, thus being in harmony with the marque's logo. By contrast, on the 4/44 the instrument group was on the car centre-line and part of a wood veneer assembly extending across the compartment and typically English in style. In the actual trim, there were changes in the design and shape of the seats, the door panels, and in the upholstery material used. These proposals, which would also apply in principle to the six-passenger

(Above) The Wolseley 4/44. The pattern of the upholstery can just be seen in this view; this differed from that used for the seats of the M.G. Magnette.
(Right) A mock-up of the M.G. Magnette version.

body, were acceptable to the Board. Whilst savouring of badge engineering, they entailed sufficient change in the external shape and interior design of each model to preserve its character. I was given the go-ahead to complete lay-out work, and build full-scale wood models and seating bucks for both four- and six-passenger cars. In this work I have to confess I was influenced by the contemporary styling of the great Italian coachbuilders - the haute couturiers of the automobile and artists of no mean calibre. I had particularly noted the black Pinin Farina two-door coupé on the Bentley stand at the 1948 Paris Salon. My aim was to produce something of this flavour but of a practical nature, capable of being made by production techniques of the time.

The base four-passenger car was given priority in the drawing office and in prototype construction. The M.G.'s character was that of a close-coupled sports saloon capable of competition in rallies and minor racing events. It was to be powered by the B-type 1.5-litre, 60bhp twin-carburettor engine. The long and short arms of the independent front suspension were pivoted to a sub-frame which was welded to the lower dash structure, transferring bending load to the sills and the central tunnel. Coil springs with concentric telescopic dampers were used. In order to provide sharper handling and response, particularly under severe conditions, the design had extra tie rods from the wishbones to pick-up points on the body frame. At the rear of the M.G. I also put a torque arm, *à la Bugatti*, bolted to the differential casing and running forward to be pivoted on the left side of the propeller shaft tunnel. The two-inch lower build allowed less room for rear axle movement. This torque arm was designed to restrain and twist down the nose of the Magnette's axle casing so as to prevent the prop. shaft touching

Auto-Architect

(Above) Probably photographed in one of the pretty villages close to Oxford and Abingdon, this very early car has no front quarterlights, over-riders on the bumpers, or spot lamps.
(Below) By way of comparison, this 1955 L.H.D. ZA Magnette incorporates all of these features.

the tunnel in the rear seat pan when the springs were fully deflected. The design of the attachment between the axle casing and the springs permitted flexibility, by using rubber pads rather than rigid U-bolts, and the front of the leaf springs were attached to the frame using rubber bushes. I was seeking the dual benefits of the lower build and more positive control for the arm, which was also intended to curb torque reaction whilst accelerating or braking. There was no need for the torque arm on the Wolseley; this was not intended to be a sporting low-slung saloon and had more space for axle movement.

Having done the basic styling and design work of the new body shell in my small drawing office, and gained the enthusiastic approval of the directors, the project was transferred to the Pressed Steel Company office for completion of all detail work. The contract for production, including the supply of prototypes, had been awarded to that company. Quite soon after the three prototypes had reached the test area at Cowley, two of the testers, Joe Gomm and Peter Tothill, were out on the A40 Witney Road. They came over the railway and canal bridge just west of City Motors, hesitated till the straight stretch ahead seemed clear, and then accelerated away in the middle lane, as was their normal practice in those relatively traffic-free days. They were travelling at about 75mph, and passing a

One of my duties as designer was to promote sales of the cars we produced. Here I am speaking at the London launch of the Magnette.

Auto-Architect

slow car in the nearside lane, when another car coming towards them pulled out into the overtaking lane. The only answer was heavy breaking with horrifying results. Joe Gomm, who was driving, came straight back to Cowley and reported violent rear axle tramp. They told me the Magnette had become uncontrollable but had eventually stopped without hitting anything. I would not believe this until I had taken a car out and found that I could reproduce the bouncing and oscillation of the rear axle (like an Austin 7 in some conditions) every time I stamped on the brake pedal. The axle wind-up which caused the tramp resulted from an unfortunate and unforeseen combination of the braking torque input, which was fed into the arm, and the particular spring rates that had been chosen.

I could not affect a cure whilst retaining the torque arm and it had to be deleted. The axle had to be attached to the springs with rigid U-bolts, as on the Wolseley. To control the increased axle movement and avoid contact between

Not all official publicity events were boring as evidenced by this picture taken a few years earlier when John Thornley (centre) and I were interviewed at the Abingdon plant by David Martin of the B.B.C. for the programme, 'Around and About'. The occasion was the launch of the TD Midget. Actually I had very little input into the design of this car as most of the work had already been completed by the time I rejoined the company. I remember there was trouble with scuttle shake on early examples and one of the section leaders in the Drawing Office under my control came up with the idea of the bracing hoop welded to the chassis which effected a permanent cure.

Auto-Architect

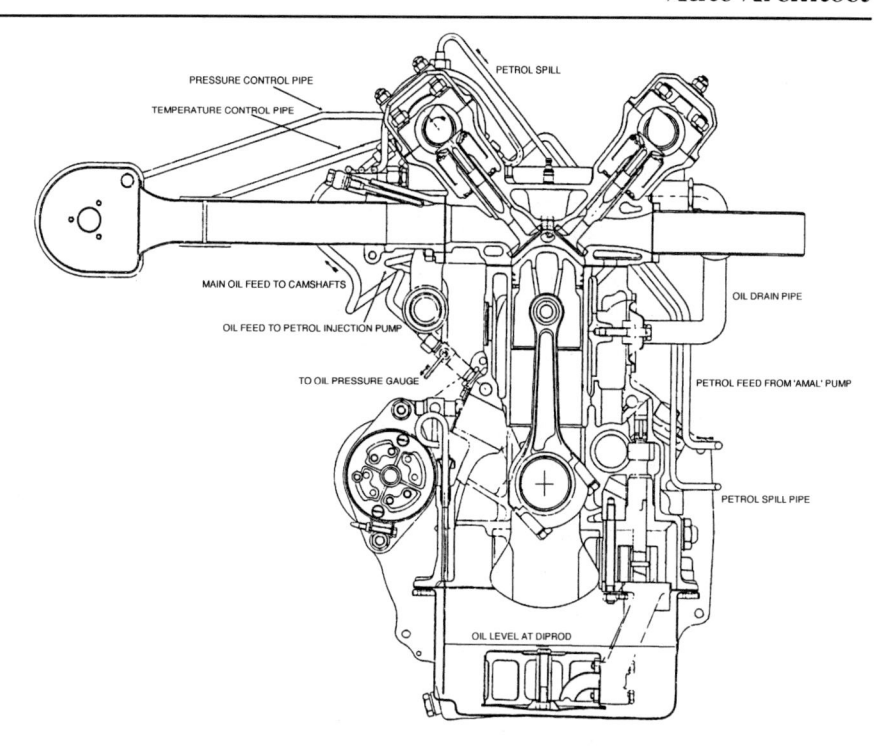

My drawing for a fuel injected twin overhead camshaft head for the B.M.C. 1,489cc B-series engine. Development of what eventually became the 1,588cc MGA Twin Cam engine was in the hands of Morris Engines Branch. It was gratifying that a supercharged unit was used in the Abingdon-built record breaker, EX181.

75

Auto-Architect

prop. shaft and tunnel, we mounted a rubber cone on the axle with a metal protuberance directly above welded to the frame, replacing the previous single bump stop. The springs also needed more camber which is why the rear of the Magnette looks higher than I had intended.

As development proceeded, I had also realised from weight estimates and engine output that the performance of the car would be disappointing. Speculating on how this could be improved, and as a line of future development not only for the Magnette but for the open two-seat sports car, I hit on the idea of doing a twin-overhead camshaft version of the B-type engine. An engine of this type would enhance the prestige and competitiveness of the M.G. marque. It would mean replacing the standard cylinder head with one having hemispherical combustion chambers. Their shape enabled larger valves to be used, giving increased gas flow and thus greater horse power. By eliminating the push rods and valve rockers, and enabling direct actuation of the valves by the tappets, the valve inertia forces would be reduced allowing higher engine speeds and still greater power to be produced. I guessed that a power output of about 100bhp could

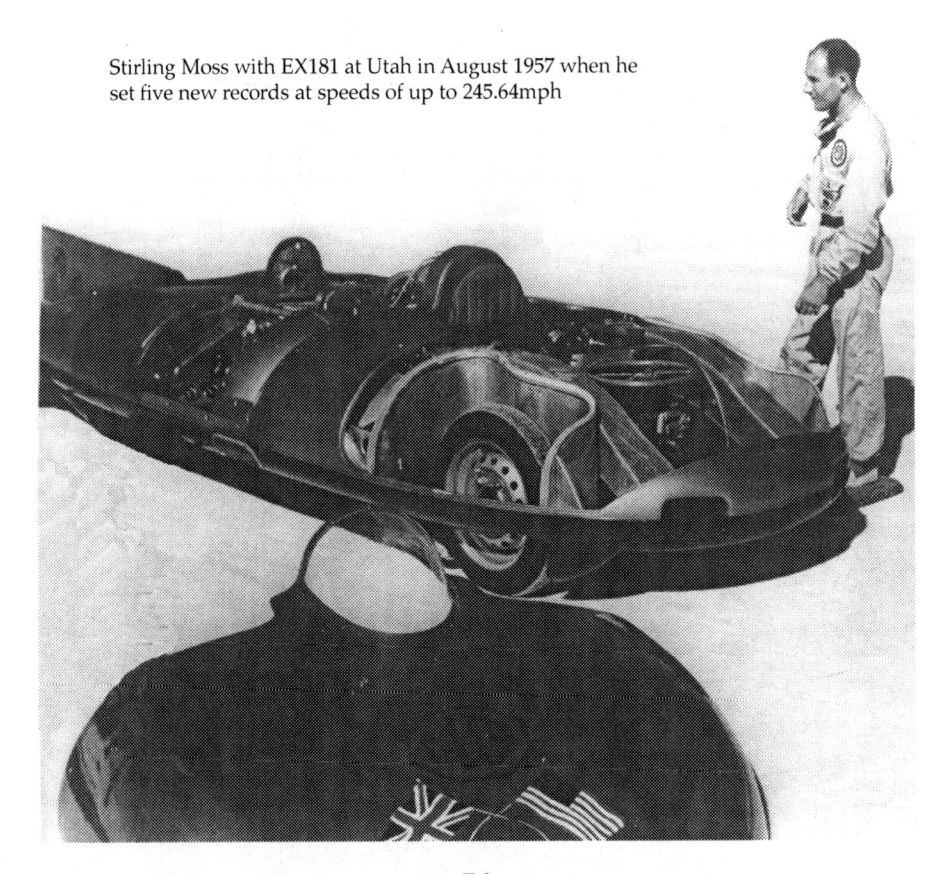

Stirling Moss with EX181 at Utah in August 1957 when he set five new records at speeds of up to 245.64mph

be achieved at about 6,000rpm. The two camshafts mounted in the aluminium cylinder head would be driven by a combination of gears and chains with a new front timing case. The standard B-type cylinder block would be retained. I then put the proposal, plus some layout drawings I had done, to my boss, Vic Oak. He agreed it was a sensible policy and authorised the transfer of the work of detail design to Morris Engines Branch under the direction of Jimmy Thomson and Eddie Maher, an ex-Riley development engineer.

Prototype engines were built and on initial testing produced 107bhp at 6,200rpm compared with the 72bhp of the standard unit. Abingdon then decided to return to record breaking. A supercharged twin-cam produced no less than 300bhp and was fitted to EX181. This streamlined aluminium machine, christened the Roaring Raindrop and driven by Stirling Moss, broke records at speeds of up to 245mph at Bonneville in August 1957. This did my ego a lot of good. In October 1959 Phil Hill broke the 250mph barrier at 254.91mph. Three years after I had left, these engines were used in the twin-cam version of the MGA sports car. I was then concentrating on my work at Vauxhall and have never fully understood why the piston problem which bedevilled production units was not mastered before that car was put on sale. Had this engine been properly developed and then fitted in the Magnette it would have given the car a performance equal to, or better than, the Ford Lotus-Cortina where a similar policy had been successfully adopted.

Although work on the saloon cars was my main preoccupation at Cowley, I was also keen to design a replacement for the aging T-series MGs. My idea was to build two versions of the same car, The 'modern' car is shown above and 'traditional' on page 79.

Auto-Architect

The years 1949-1955 were a period of great activity for me, during which I was engaged on a number of design projects, some of which came to fruition and others which did not, but all were interesting and taxed my creative ability. I had never lost my first love of the racing car and its open two-seat sports car offspring; this represented the cutting edge of my profession as an automobile engineer and designer. At the time I had given birth to, I hoped, worthy future closed car models. But the design of the open two-seater M.G. had eluded my ambition, remaining in the capable hands of Syd Enever at Abingdon. These were orthodox in construction, having a chassis with substantial box section members upon which were mounted a wood-framed, steel-panelled body made by Morris Bodies Branch in Coventry. The whole assembly of body, chassis and mechanical units came together on the track at Abingdon. The attractive styling of the car was

These pictures of the MG sports car design show the side chrome moulding and a possible layout for the dashboard. Front suspension used torsion bars and the power unit would have been the XPAG/XPEG unit from the T-series model.

Auto-Architect

The 'traditional' version of the projected new MG sports car had flowing wings, rear hinged doors, wire wheels and folding windscreen, all features taken from the existing model and thought to appeal to American customers.

as English as roast beef, with typical sweeping front wings and the spare wheel mounted externally at the rear. It was at this time that there appeared an Italian styling exercise by Bertone on an Alfa Romeo chassis, named the Disco Volante, a most attractive all-enveloping style which set a debate going on which way M.G. should go - remain roast beef or go cannelloni? My own opinion was that the US market would favour the English styling and other markets the Italian or both. Would it be possible, I wondered, to offer two versions of the same car by having

Auto-Architect

a basic sheet metal carcass, or structure, which would be common to both, and fix to it the external sheet metal panels, namely front and rear wings and door panels, which determine the type of styling? I did some quick layout work and this seemed feasible, so I had two full scale wood models built of each version. They were attractive models, all agreed, and I was given permission to build a prototype, which I had done to the English version. It was considerably lighter than the then current TD model, so with the twin-cam B-type engine M.G. would have acquired a power unit in keeping with its character and provenance, and have great scope for future development. However, there was divided opinion on this approach, and as the Abingdon team had, in parallel, produced the attractive MGA prototype of orthodox safe construction using proved production methods, this was eventually chosen as the successor to the TF.

Chapter Nine

Pathfinder and Six/Ninety for B.M.C.

The Riley (which was to be named the Pathfinder) was similar in character to the M.G., being a low built sports saloon, but having six-passenger accommodation. There was considerable difference of opinion on this among the ex-Riley executives at Cowley and Abingdon, some of whom envisaged the new Riley as a four-seat car, despite its main competitor being the large Jaguar Mk VII, and the use of the same body for the large model in the Wolseley range.

I took the occasion of a visit to the Geneva Motor Show to question the Swiss agents on this matter of size and all were in favour of a six-passenger car to compete with the Jaguars, Daimlers, Mercedes and others. With the 2.5-litre four-cylinder hemispherical head, 110bhp engine, I envisaged this car as a true *voiture de grande tourisme* as the French would describe it.

Looking into the future, as a line of possible development after the phasing out of the Riley engine, I had in mind the development of a twin-overhead-camshaft version of the 3-litre, six-cylinder C-type engine in the same way as had already been started for the B-type. I believed a reliable unit could be made with fewer potential problems than with the high output smaller engine. Such a car could have been serious competition for the Jaguars which still retained their old fashioned leaf spring rear suspension. This dream was but a gleam in my eye and never became a reality in my time. However, I later learnt that after I left one engine was built.

In many ways the Riley was the most difficult design exercise in the four car programme. This was because the RMF model it was to displace was a very good and much loved car. Certain of its features were associated with its good qualities in the minds of surviving Riley people who believed that they should not be abandoned in the new model. Principal among these was that it should have a robust chassis frame upon which the body was mounted, and that the rear axle should be of the torque-tube type giving better axle control. With both of these recommendations I was in agreement. With a combined engine-gearbox weight of about 800lbs I was hesitant to mount this in a light sheet-metal structure. So far as the axle was concerned, there was no choice – I had to use the standard

Auto-Architect

BMC hypoid type. However, I was able to achieve the virtues of the torque-tube by copying an arrangement used by Buick in the USA. This consisted of using a rigid axle with an open propeller shaft, but providing two diagonal radius arms. Each of these was fixed to the axle beam near the wheel hub and at their forward

(Above) 1950 mock-up of the Riley Pathfinder with 'V' windscreen, prominent chromed grilles either side of the radiator shell and quarter-lights in the doors. (Below) A sketch showing the rear suspension layout.

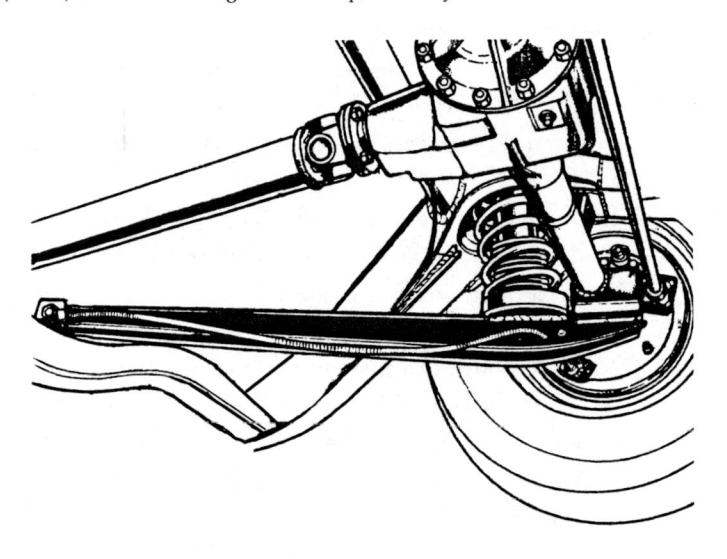

Auto-Architect

(Above) A Riley Pathfinder chassis in the experimental workshop at Cowley. The coil-sprung rear suspension and radius arms can be seen, as can the rear-mounted servo and 4-cylinder, 2,443cc Riley engine. The perimeter chassis frame with its deep side members has the cross bracing placed so as to give sufficient room for the foot wells, thus making possible a low seating position.
(Below) An early publicity photograph.

Auto-Architect

ends were resiliently pivoted through rubber blocks to the frame crossmember, as close to the forward universal joint as possible. Transverse location of the axle beam was by a so-called panhard rod, pivoted at one end to the axle and at the

The Riley Pathfinder was announced in October 1953 but deliveries to customers did not really start until well into 1954. The 2,443cc, twin-camshaft, OHV Riley engine produced 110bhp, giving the car a top speed of nearly 100mph.

other to a bracket welded to the opposite frame sidemember. Coil springs with telescopic dampers provided the suspension medium. At the front conventional independent suspension was used, with torsion bar springs splined into the lower arms and anchored to a frame crossmember amidships.

The frame was of the peripheral type to secure the lowest possible floor level in the rear compartment - a feature often used later in US cars. One feature I was not able to adopt was rack and pinion steering, due to having to make provision for the longer C-type engine which interfered with the rack. The cam and lever alternative was always a little spongy.

The three prototypes built to this formula all performed well with no major disasters that I can recall. In fact they were so promising that following a test drive I gave Vic Oak and George Dono (director of Nuffield Metal Products), a decision was made to go ahead with the project. The frames were to be made by John Thompson at Wolverhampton, with the body built at NMP in Coventry, to whom the sheet metal detailing work was transferred. The components would then be assembled at Abingdon. Therefore, it was with great concern and surprise that I learned (when I was at Vauxhall) of the disasters that occurred to some early

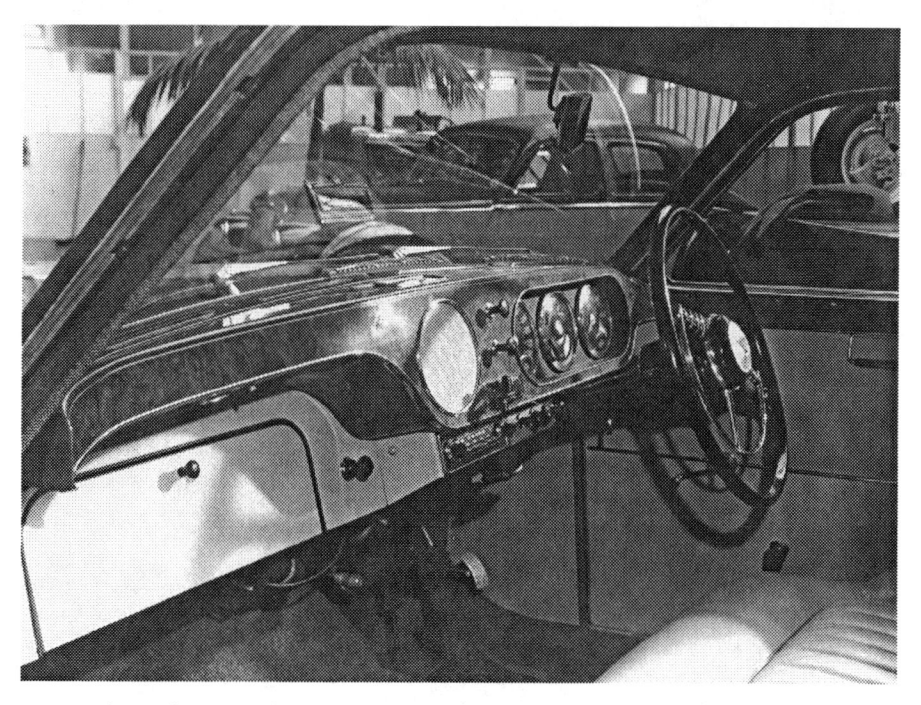

This dashboard is the early version and the design was changed in 1956 for a full-width, walnut veneered panel. The right-hand gear lever was mated to a BMC, four-speed gearbox with synchromesh on all but first gear. An overdrive later became an option and this gave very relaxed gearing of 28.71mph/ 1,000revs in the highest ratio.

Auto-Architect

production cars due to loss of rear axle control caused by failure of the panhard rod frame bracket. Had such a major failure occurred on prototypes, remedial action would have taken place to reinforce the anchor point by whatever means were necessary. I am sure these disastrous failures were caused by poor quality or inadequate welding on early production frames. A service campaign to overcome them was necessary and carried out. This included the recommendation to fit retrospectively the additional tie-rod to early cars that was introduced in production at car RMH 3498. This was bolted to a new lug added to the panhard rod bracket, and ran across the car to be bolted to another new U-shaped bracket welded underneath the opposite frame member. I do not reckon this extra strength (it wasn't another panhard rod) was strictly necessary, provided the welding on the panhard rod bracket was sound, or had been correctly rewelded. However, as a fail-safe addition it certainly did no harm.

On the choice of the braking system one must remember that at this time rapid development on disc brakes was taking place, but mainly by the Dunlop Company who were not a supplier to BMC. It was with considerable interest, therefore, that I had a visit from Joe Kinchin, who was chief engineer at Girling, to demonstrate their parallel development to the disc – the two trailing shoe drum brake. The main advantage of this was that it was free of the pronounced self-

Like the Magnette, the Riley Pathfinder gained swivelling quarter-lights in the front doors once production was under way. The twin fog lamps are built-in.

servo effect of the two leading shoe drum brake in the forward direction, and less sensitive to lining debris and thus fade, so that its performance was more constant with changing atmospheric conditions and lining characteristics. Furthermore, manufacturing processes were well established and there would be no delay in production. Like the disc, however, it was entirely dependent on a reliable pneumatic servo for reasonable pedal pressures. I was impressed by the technical arguments in favour of this system, and the convincing demonstrations of it. After consultation with my superiors, and equipping one of the prototypes with a trial system which gave satisfactory results, I decided to adopt it for production.

PATHFINDER'S POTENTIAL

The Autocar's August 1956 Road Test of a later Pathfinder, still with the rear coils and in this case fitted with overdrive, underlines the car's ability:

'The combination of long torsion bars and coil springs give a suspension suitably firm for high speed driving and yet quite comfortable when taken slowly over indifferent surfaces. The driver can choose a line with complete confidence. No tendency to wander at any speed. A feeling of being wafted along in comparative silence with speedometer on 90mph mark and rev counter recording 3,000rpm.'

David Rowlands who produces a Pathfinder Newsletter for the Riley RM Club writes: 'I have had Pathfinders since 1961. My forty-two-year-old un-restored example is fast, smooth, utterly reliable and extremely comfortable. In my view Gerald's rear suspension is inspired. It gives a supple, soft ride which is adept at smoothing out surface fluctuations. The Pathfinder is in everyday use and will still cruise happily all day at 70mph with plenty of power in reserve.'

David also writes of the current Australian scene where, at the last count, there were over twenty cars in use in Victoria and others scattered around the continent. Brian Jackson of Queensland has had his since 1955. It is referred to as 'The fast smooth Pathfinder - God's gift to automotive man!'

Ken McKay in Western Australia now has the car in which Alan Howie steadily clocked up 20,000 miles a year into the 1980s. Alan used to regularly drive the 2,500 miles from Perth to Canberra in two days, stopping only at Ceduna, and cruising between 75 and 80mph. The car has now covered over half-a-million miles on the original engine, which has been re-sleeved and fitted with new pistons.

There was, however, one aspect which was disturbing - our lack of knowledge and experience of servo systems and for this we relied completely on Girling. The size of the servo required to give an acceptable pedal pressure was such that it could not be housed in the engine compartment as is usual. Due to the late stage in development the only space available was bolted to the main frame under the rear seat pan. This position, despite requiring rather long hydraulic fluid and air pipes, was deemed satisfactory and was adopted. It was also unduly exposed to the elements and it may be this factor, perhaps caused by corrosion or even physical damage due to flying road debris, which gave rise to a very indifferent service picture, although specific examples of brake failure were difficult to find. Regrettably service problems of varying degrees of severity occurred on this model throughout its production life, many of which remained unsolved. In the attempt to continue the unusual features which would appeal to discerning Riley owners, they were not very well developed.

RILEY CHASSIS FRAME WELDING AT WOLVERHAMPTON

In 1997 Peter Tothill recalled the early days of 6/90 assembly at Cowley where he was production engineer. 'Too often the N.M.P. body just did not fit when it was dropped onto the Thompson frame on the moving line. The number of recalcitrant units which had to be taken off the line for attention in the rectification area multiplied and we consulted our colleagues who were building the Pathfinder at Abingdon. We explained that we had too many miscreants which had to be bashed together using hammer and oxy-acetylene and Abingdon confirmed that they had similar difficulties. However, it was less of a problem for them as each Pathfinder was built at a separate work station so that were was no moving line to disrupt.'

After further consultation Management decided to travel to Wolverhampton where John Thompson were meant to be welding on a special jig supplied to them by Nuffield. On a tour of inspection the visitors found to their consternation that these early 6/90 and Pathfinder frames were being assembled and welded whilst propped up on on wooden trestles with not a jig in sight!

After some acrimony between the two managements the matter was sorted out and frame production did continue, this time using the correct jig. Gerald's conjecture about inadequate welding is thus proved correct. It is further confirmed by those who have totally rebuilt early cars in Australia and found evidence of poor quality or inaccurate work.

Auto-Architect

The transformation of the four-seater and six-passenger body shells into versions which were acceptable as Wolseley models (by the changes already described) entailed a considerable amount of drawing office work, which my small office was unable to handle, so some of this was delegated to the main Morris office. In the case of the 4/44, to suit phasing out of the former engine, it was planned to make a change from the Morris XPAG to the BMC B-type early in its production run, thus entailing several minor installation changes. In the case of the 6/90, alterations of a similar nature were required to accommodate the installation of the C-type engine. Additionally on this model it was decided to use the Lockheed brake equipment from the previous 6/80 model, and to adopt a steering column gear change in place of the right-hand, floor-mounted lever I had chosen for the Riley.

All these changes were far removed from what was derisively called badge engineering by some critics, and together with conservative seat trim and instrument panel styling, established for the cars a distinctive Wolseley image.

At Abingdon assembly of the ZA Magnette was carried out on a line adjacent to the MGA 1500 two-seater once this was introduced. The Riley Pathfinder was also assembled in the same building on a line just beyond the pillars to the left of the picture.

Auto-Architect

(Above) The earlier ZAs had this dashboard which had a metal capping that was painted to resemble veneered wood. There were two ashtrays, one on each side of the lower panel.

(Below) The ZB and Varitone layout. The later ZAs also had this dashboard but with a flat, not dished, steering wheel and a central ash tray on the top panel.

Auto-Architect

When the ZB model arrived in 1956 it was easily identified from the ZA by the straight chrome strips on the front wings. By then the engine developed more power and the axle ratio was raised to provide more relaxed high-speed travel.

When the ZB was introduced in 1956 the Varitone joined the range. On this model the rear window was enlarged and two-tone paintwork introduced, this had a chromed strip at the waist line to divide the two colours. Some Varitones were finished in just one colour but the chrome dividing strip was retained.

Auto-Architect

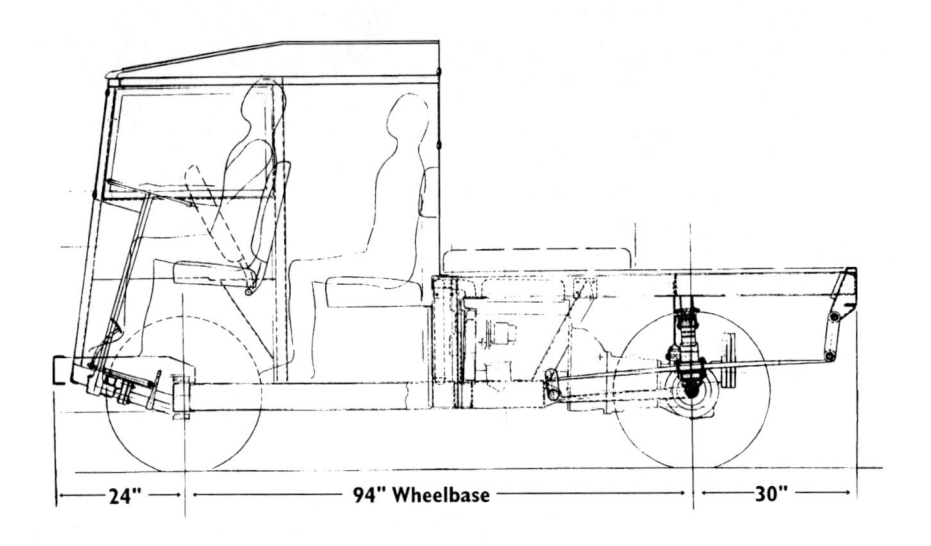

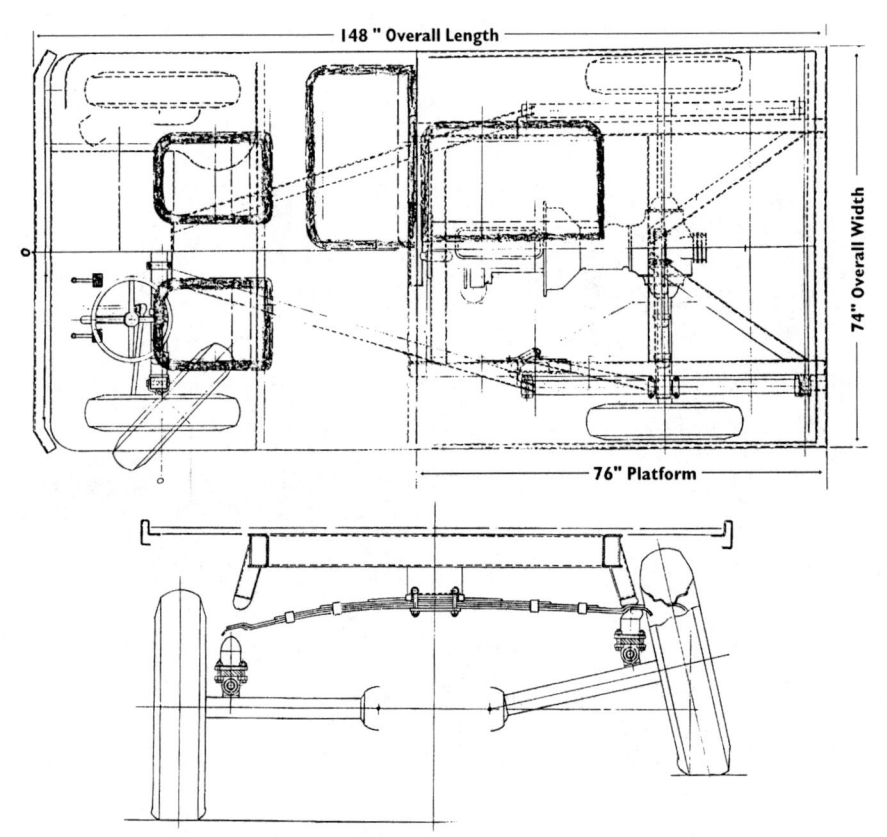

Yeoman side elevation, plan view, and rear suspension details from my drawings for the Perkins installation.

Chapter Ten

Wansbrough's Yeoman and Orchard House

Another design project came from George Wansbrough who visited me in Oxford in late 1949, by which time he had relinquished the Chairmanship of Jowett. George was of that generation of Cambridge graduates influenced by Keynesian socialism and the Attlee administration had appointed him as a Bank of England director. In that capacity he had, in 1948 whilst still with Jowett, looked at the economic situation in India with the object of reviewing Britain's policy after independence and partition. His thorough and wide-ranging report gave particular attention to the road transport and motor manufacturing industries, and of the latter it was critical. In 1945 Miles Thomas, representing Nuffield, had made an agreement with Birla Brothers in Calcutta to manufacture the current Morris Ten in India, at first by completely knocked down (CKD) assembly and finally by total local content. Whilst what then became the Hindustan Ten, though dated, was suitable for city and suburban application, it was totally unsuitable for village and rural use.

What was required, the report said, was a simple robust vehicle which could carry ten one-hundredweight sacks of rice, or six passengers, or a combination of both, i.e. a pick-up truck with a six-passenger cab. Furthermore, it should be easy to manufacture and require the minimum of tooling costs. On his return from India, and envisaging the possible use of the Jowett flat-twin engine and four-wheel drive for such a machine, George had discussed the idea with Charles Grandfield. Charles, remembering his desert experiences with Montgomery, had been emphatic that the Jeep, with its four-wheel drive capability, enjoyed little advantage over the much lighter German military VW with its rear engine, two wheel drive, and ZF limited slip differential.

What happened to the report I do not know, but George was sufficiently certain of the possibility of this specification, not only in India but in developing countries generally, that he started a project to build a prototype. He was able to privately finance this, and get the design work done by a freelance Anglo-Italian designer, Achille Sampietro. He worked in conjunction with Thomson & Taylor Ltd., well-known experimental engineers at the old Brooklands race track, with

Auto-Architect

whom he formed a company called Yeoman Vehicles Ltd. The prototype was completed but I think George was not satisfied with it, which was the reason he called on me to get my opinion. While it complied with the base specification, having seating capacity for six with a flat platform body behind the cab, the structure was complicated, 'bitty', and not easy to manufacture. The Jowett 2-cylinder power unit, mounted under the platform floor at the rear and driving forward to the rear axle, was not sufficiently powerful but was the only suitable one available at the time.

After discussing the project in general and the shortcomings of the vehicle in particular, he asked if I would undertake redesigning it, such was his faith in my ability! I was heavily involved in my new job at Cowley by then and the only possibility would be to do this as a spare time project working in conjunction with Ken Taylor of Thomson & Taylor and his one lone draughtsman. To help George I agreed to do this, and spent many late nights at my drawing board doing the basic layouts of the Yeoman until they were advanced enough to obtain agreement. At this point they were transferred to Thomson & Taylor for completion and detailing. This meant evening visits every second week. It took about an hour to drive the 40 miles to Brooklands and I used to get back to Orchard House very hungry about midnight. There wasn't time to fit in a meal.

The liaison with Thomson & Taylor was a happy one, mainly because Taylor was an experienced and practical engineer and we got on well together. Progress was rapid despite the new gearbox I designed which was found to

The Petter-engined Yeoman loaded for testing.

have one forward speed and four reverse speeds when first assembled! Luckily the fault could easily be put right. The new design externally looked very much like the first one, but in construction and detail differed considerably. I drew one version with the same Jowett engine and another to take the four-cylinder, 108 cu ins Perkins diesel. Both were mounted underfloor but now in front of the axle with a new gearbox/final drive unit, with a power take-off to permit light agricultural machinery such as ploughs, harrows, water pumps, etc. to be driven. Importantly, it had a lockable differential for use in muddy and slippery conditions, such as those encountered in farm yards.

The frame was in two parts, one virtually consisting of the load platform at the rear over the engine, and one forward at a lower level, shaped like an A in plan, to support the passengers; the two being firmly bolted together. Independent wheel suspension was provided at the front by the simple system of two transverse leaf springs bolted to the apex of the front chassis A frame, one above the other. The usual king pins for the steerable road wheels being mounted between them at each extremity. At the rear a simple swing axle system was used. The suspension medium being longitudinal, half-elliptic leaf springs trunnion-mounted to the half axle tubes and shackled directly to the load platform. A

At the Army test track at Long Valley, Aldershot the provision of two-wheel drive with a lockable differential gave adequate traction. Here George Wansbrough is at the controls.

Auto-Architect

further transverse leaf spring was bolted to the centre of the cargo platform and only came into contact with the axles with a full load. The whole concept was that of a simple, semi-agricultural vehicle in which appearance and performance were of secondary importance. With the initial exception of the power unit, it could be manufactured from readily available material with the simplest of easily made tools. There was not a single component in the whole vehicle which could not be made from flat sheet steel, sheared and flanged, except for the roof over the passenger compartment which could have been a fibreglass moulding. So no elaborate press tools were required, keeping costs to a minimum.

This design was not built with the Perkins engine. Thomson and Taylor modified the second prototype by reverting to the layout of the original final drive unit and mounting the air cooled PAV-4 Petter engine behind the rear axle in the tail of the machine. They also modified the cab with central steering and just one front seat. Both my frame and my suspension design was retained. A programme of testing on and off the road was then carried out to reveal any defects. This included some use on a farm, and a particularly tough session on the muddy Army test track at Long Valley, Aldershot. Its good performance there confirmed the opinion that a heavily loaded two-wheel-drive rear axle (75% gross vehicle weight) with a lockable differential is equal to a four-wheel-drive vehicle like the Austin Champ or the Land Rover.

The basic cab was designed to carry up to six people with the canvas roof giving them some protection from the elements.

It was never part of George Wansbrough's project to manufacture this vehicle himself, but rather to demonstrate that it could be commercially viable and bring benefit to developing countries. Therefore, he started a series of visits to manufacturers in the UK whom he thought might be interested in producing such a product. Despite great perseverance on his part he did not succeed. Either the timing in 1953 was not right, as most firms by this time were well advanced with their own post-war programmes, or they were unwilling to risk entering a new, and to them untried, market. One ray of hope came in the 1970s when it attracted an enquiry from the White Motor Corporation of the USA. I had by then retired from Vauxhall and George and I mounted a presentation in his suite at Claridges for their well-known president, Bunky Knudson. While they were interested in and praised the vehicle, they ultimately turned it down as being insufficiently powerful, which was a general criticism of which we were aware. This was particularly disappointing and frustrating for George who perforce had to abandon the project for financial reasons.

Looking back, one sees that time and events overtook the Yeoman and other similar projects, including United Nations ideas, to produce a simple vehicle for developing countries. The larger manufacturers saw the need and developed the car based pick-up truck which satisfied most of the requirements and was cheaper. But they did not bring motor manufacturing industry to the smaller countries which George had hoped to help.

Orchard House was constructed in that post-war period when the limit in force on using building materials meant that the floor area could not exceed 1500sq. ft. I am happy to still be living there, although selling the cottage at Iffley (page 98) was a wrench at the time.

Auto-Architect

Then there was the commissioning of Orchard House. Soon after returning from Yorkshire, Diana had found a derelict market garden off Tree Lane in Iffley, near our cottage which had been let whilst we were away. We were friendly with an Oxford architect, Albert Harvey, who wanted to create a modern design. Together, we plotted a building facing south with space and light in the main rooms. The curving staircase, cream terrazzo on an iron frame and inspired by one at the Alfa Romeo head office, was fixed directly to the curved outside wall with no other support. It is still in place 45 years on! There weren't today's planning restrictions, but there were limits on expenditure and floor area - not more than 1,500 square feet. I have not kept the bills but the total cost must have been around £4,000. We didn't have the capital, so the cottage was sold to provide funds. During the year whilst the house was being built, we stayed at first at a hotel by the river at Dorchester run by Paul Griffiths (who worked at Lucas) and his wife, and then moved to a flat in North Oxford. Even five years after the war there were difficulties and delays in obtaining the materials but in the end we have all benefited from the effort involved.

It was also that year that we finally sold the Deroy. There are moments when I have regretted this but, with all the new work at Cowley and the need for mobility until the house was completed, there wasn't time for everything. As it had served its purpose, it seemed sensible to sell it to the South West Region representative for Jowett, a man called Hodgson, who was keen to take it on. The last I saw of the Deroy was it proceeding smoothly up Oxford High Street. I have wondered whether the remains still lie in a shed somewhere. There might be some historical interest in that rear suspension design now that independent arrangements have been so widely adopted.

Chapter Eleven

Chief Engineer and the Clash with Leonard Lord

It was in June 1952 that Alec Issigonis asked me one day what it was like to work for a small company. I was somewhat surprised as small cars, which were his speciality, needed to be produced in the large quantities which were beyond the resources of a small firm. I told him what it was like as best as I could, stressing that you were your own boss! To my, and I believe everyone else's astonishment, a few days later he resigned from Morris Motors and joined the small Alvis Company in Coventry, presumably to design them a new product.

The departure of Issigonis from Cowley caused a hiatus in the senior mechanical personnel of the company. Vic Oak, technical director to whom both Alec Issigonis and I reported, decided to retire for health reasons, as did John Rix, who was technical director of Austin. This left me as the most senior executive with wide original design experience left in the company. I was promoted to be a local director of Morris Motors Limited in charge of technical matters and reporting to R.F. Hanks, the chairman. At the same time I was made Chief Engineer - Chassis & Body - of the British Motor Corporation, so that I had a dual role to fulfil at Austin and Morris.

This meant dividing my time judicially between the two plants to chair or attend meetings concerning future and current engineering problems. It was 60 odd miles between Cowley and Longbridge and on a good day took an hour and twenty minutes. At least it was a chance to think and I was experiencing my own designs in typical conditions. I did my bit to bring the two teams together! As a day out from their duties in the spectacular Exhibition Halls, BMC engineers attending the Frankfurt Show were invited by me to lunch at a restaurant by the river Rhine at Assmannshausen. Relationships were nurtured whilst enjoying Rhine salmon and pink champagne and some of the Longbridge people told me afterwards that they had never had that type of invitation before.

Despite my woeful lack of managerial experience, I came to realise that my task had echoes of a Prime Minister in Cabinet. I had to make decisions having been advised of the details by experienced people who had coped with the job for years. Many of the current problems related to reports received from

Auto-Architect

the service department. There were still worries about models which had been designed before my time. In Canada there were piston failures in the side-valve Morris Oxford engine in which the traditional grub screw had been used to hold the gudgeon pin on the connecting rod. Because of metal shrinkage in the extreme cold the gudgeon pins were trying to rotate, and it was necessary to change to the newer method of floating pins secured by Seeger circlips. Then we established that the location of the water pump, fitted on the hot rather than the cold side of the cooling system, was contributing to the overheating of the Morris Six and Wolseley 6/80 ohc engine. Of the, then, current cars the Pathfinder continued to harass us. Apart from the already mentioned mechanical troubles, water was leaking into the bodyshell that it shared with the 6/90 and we found there were deficiencies in both the door seals and the apertures for the windscreen wipers.

The car that Lord cited as a pretext for sacking me was the Wolseley 6/90. This was introduced in 1954 and instad of the old Wolseley 6/80 unit was given the 2,693cc, six-cylinder BMC engine which produced 95bhp. The chassis and suspension was the same as the Riley but the Wolseley had a steering column gear lever. The Wolseley sat 2in higher than the Riley and there were changes to the body around the sills and wheel-arches. A Series ll Wolseley 6/90 was introduced in 1956 which had the Riley right-hand gear lever and a revised facia. Both the Riley and Wolseley were later given half-elliptic rear suspension.

However, it wasn't all office work and committees. Bertie Kensington-Moir was involved with a London Nuffield agency and when he lunched at Cowley to discuss the Metropolitan Police order for 6/90s he regaled us with his memories of the Bentley boys. Another enjoyable day out led us to Harry Ferguson's Cotswold mansion, Abbotswood. After lunch we were taken to watch Tony Rolt's unsuccessful attempts to drive a variety of conventional cars across a very muddy field. He then mounted the crude four-wheel-drive Ferguson prototype, with its third differential and Salerni couplings, sailed through the mud patch and continued across the grass at an impressive speed. The irony of the day came from my discovery that the machine used a Javelin engine and gearbox.

As to the future, it was difficult to plan ahead when there could be overriding orders from Chairman Lord and his henchmen. There was no forewarning of what Lord had in mind. An example was Dick Burzi's clay model for a replacement Minor which was sent down from Longbridge with Lord's approval. I was taken round to the sheds where ideas were hatched and told that this was to be the new car. It was just a restyled, full-width body to the same configuration. There seemed no attempt at a more imaginative use of space to give increased passenger accommodation. I wondered about suspension developments which we at Cowley believed were essential for world markets.

I was made more aware of the tussles ahead and the difference between the philosophies of the Austin and Morris camps. Due to my long association I was more at home with Cowley people than I was with those at Longbridge now in command.

I need not have worried. My term of office in the company soon came to an end. An otherwise enthusiastic *Autocar* road test in September 1955 of the new Wolseley 6/90 had criticised some of the car's features; sponginess in the gear lever, insufficient left leg room for the driver, reflection in the oil gauge glass, and high brake pedal pressures. Leonard Lord had noted these unfavourable comments. The three faults could soon have been corrected and Lockheed were aware of, and working on, the acknowledged brake problem. However, Lord regarded this criticism of the Wolseley, together with the adverse service reports and high warranty costs on the Riley Pathfinder, as my responsibility. The call came to visit Longbridge. I stayed at Droitwich before driving over in a Magnette for a morning interview. I had been summoned by George Harriman, number one henchman, charming, but still a 'yes-man'. Handing over the pages torn from the magazine with the offending paragraphs personally marked in orange crayon by Lord himself, Harriman told me that the chairman was not pleased with the test comments. He then said that Lord demanded my resignation from the position I held or, alternatively, my demotion to a lesser position in charge of the development of gas turbine cars. As I had neither knowledge of, nor faith in, the latter, I had no hesitation in accepting resignation, trusting that I would find a position to continue my career in the industry in a less turbulent environment. At no time then or later did I have any personal contact with Lord.

Returning to Cowley I remember thinking mainly of self-preservation. I was out of a job aged 44 with a wife, small daughter, a house we wanted to keep, and little capital. Kindly Wilfred Hobbs, Lord Nuffield's former assistant and still a director, came to my office and advised me to clear my things that evening. Lord was visiting the following day and it would be best if I was out of the way. I drove across to Iffley and reported to Diana that the husband who had left the previous day as a director of a large company (a position I was never again to achieve) was now unemployed. Always aware of the ups and downs of industrial life, Diana was a model of stoicism and comfort. Friends were sympathetic and I was particularly grateful to Max Goldsmith, chairman of Metalastik, who came over from Leicester. Reggie Hanks, always solid and dependable, bothered to get in touch and spoke about me to Patrick Hennessy at Ford. Had he then known his days were numbered? Two weeks later his own resignation was announced. Then, in November, the weeklies reported that the Alvis project had been abandoned and that Alec Issigonis was to join BMC.

Issigonis had kindly invited me to see the prototype at the Alvis works. My vague recollection is that it was not particularly striking in appearance, rather like the then current Morris Oxford Series II and not likely to appeal to traditional Alvis clients. It was powered by a V8 engine, but its outstanding feature was undoubtedly four-wheel independent suspension of the Hydrolastic type, at the time being developed by Alex Moulton and Issigonis, and which would later be adopted by BMC.

Had I continued as BMC's chief engineer I would have hesitated to adopt the Hydrolastic concept. If it was so superior, why didn't other manufacturers, like Fiat, Peugeot and Renault follow that route? I was later to experience the 1800 saloon. The car certainly had good roadholding and masses of room. On the debit side, I thought the ride was too choppy, the driving position uncomfortable, and the steering awkwardly cumbersome at slow speeds, features which compared unfavourably with the competition. Before he departed for Alvis, Issigonis was working on a power unit comprising an A-type engine with a special oil sump housing a four-speed transmission with differential and drive shafts. If the unit was mounted transversely the shafts could drive the front wheels of a small car. I immediately connected this with his obsession with a small 'charwoman's car' about which he had talked, and which I have previously mentioned. Was this the power unit? The project was left in the capable hands of Jack Daniels, an old M.G. man who had become chief experimental project engineer under Issigonis.

HANKS, HARRIMAN, AND HOBBS

Gerald speaks well of Reg Hanks and one of the many unanswered questions is what would have happened (and what designs would Gerald later have produced) if the Nuffield Group had continued as a separate organisation under Hanks' leadership. He had been with Morris since 1922, following war service and engineering training at the Swindon Railway workshops. Appointed Managing Director in 1947 he pushed through an agreed plan for centralised control at Cowley, abolishing autonomous units. There are records of an urbane, steady character who knew where he was going. He accelerated the production of the Minor and opposed the merger with Austin. Although he became one of the two deputy managing directors to Leonard Lord when the British Motor Corporation was formed in 1952, his tenure only lasted until 1955 when he returned to his roots as chairman of the Western Area Board of British Rail.

George Harriman (Sir George in 1965) was the other deputy to Lord with whom he had worked for most of the previous thirty years. He had been at Morris and then went to Longbridge as machine shop superintendent when Lord was revitalising Austin. By 1945 he was general works manager and a director. Unlike Hanks (he was twelve years younger), it is recorded that he did not stand up to Lord. However, he was able to keep B.M.C. going, after Lord's semi-retirement in 1961, until the British Leyland merger in 1968. He was a thoughtful and reasoning character known to be mindful of other people's feelings and would not have enjoyed the September 1955 interview with Gerald.

Miles Thomas (Reg Hanks' predecessor) wrote in his autobiography that Wilfred Hobbs 'Played an almighty part in the development of Nuffield.' A bachelor accountant, who had been articled to Price Waterhouse, he was appointed secretary to Morris Commercial in 1924. Subsequently based at Cowley be became one of William Morris' chief assistants and a loyal friend. It was Hobbs who brokered the disagreement with Lord in the 30s and he later accompanied the older Lord Nuffield on his sea voyages to Australia. In 1946 he became Secretary of Nuffield Products. It was entirely in character that he should have had the last contact with Gerald. The conversation is not remembered but Gerald says that Wilfred Hobbs, no doubt himself concerned by the course of events, did manage some words of comfort.

Auto-Architect

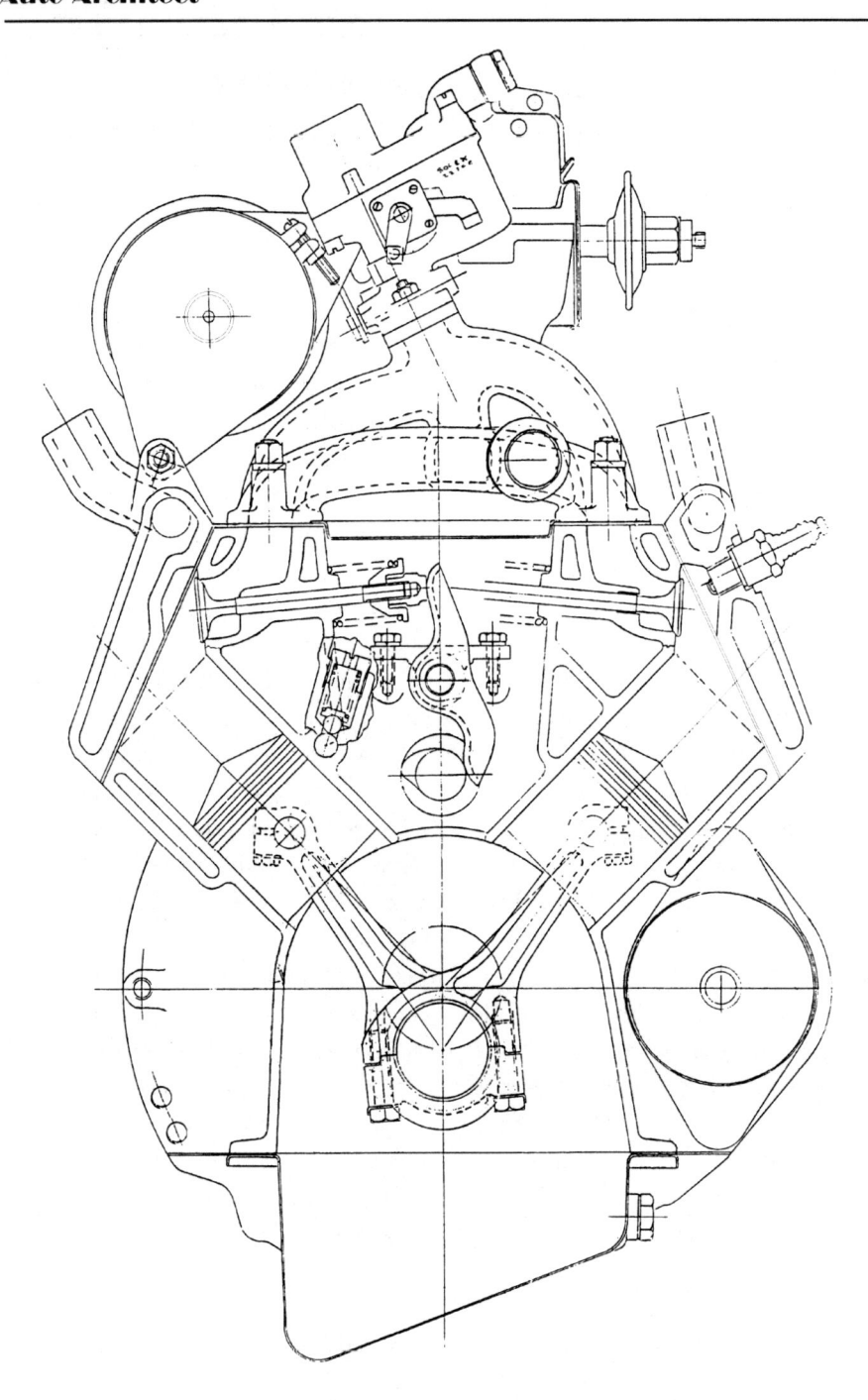

My proposed simple and cheap-to-build V4 engine for possible use in a BMC small front-wheel-drive saloon.

In due course it was mounted transversely at the front of a standard Morris Minor bodyshell, driving the front wheels, and thus became the prototype of all future BMC front-wheel drive cars.

There were, therefore, two identical cars for comparison, one with conventional rear-wheel drive and one with front-wheel drive. On normal level roads there was little to choose between them, but on icy roads and snow covered hills the front-wheel drive car was markedly superior. It was able to steadily ascend hills which defeated all rear drive cars. I would have liked to continue along this path producing front drive vehicles with long travel suspension and a development of the beam rear axle which had impressed Lanchester before the war.

At the time of these tests there were no cost figures available for these two concepts, but I had a gut feeling that the front drive car was the more expensive of the two. In contemplating how to reduce this, it occurred to me that some of the cost would be in the engine, and that a simpler V4 could be used which at the same time would be more compact. During my time at Jowett I had given some thought to this type of engine for the Javelin and had done a few sketches, but I had discarded it for various reasons. I now realised it might be suitable for

---

THE MORRIS MINOR REPLACEMENT

The proposed replacement, which Gerald saw in model form before he left, never became a Minor. In a marketing move praised at the time, the new style was transformed into the svelte compact Wolseley 1500 and the performance-orientated Riley 1.5. But the British Motor Corporation management was not coming to terms with what was happening in Europe. Productions figures above 100,000 may have seemed satisfactory at the time, but whilst B.M.C. were working on so many different models, Peugeot's factory space was devoted to the 403 and 404 and made well over a million of both of them. That was the way to recoup development costs. Gerald's vision of one new, good vehicle, competitive in all departments, and not just using an existing engine because it was more convenient, was so badly needed.

This Wolseley-Riley combination was not the end of this style. It was taken up by B.M.C. Australia for manufacture in their new factory at Victoria Park, Sydney. The resulting Austin Lancer and Morris Major were not popular. Customers still preferred the Volkswagen with just as much rear seat room, much better suspension and proven reliability. Would there have been a different story to tell if Victoria Park had produced a developed version of Gerald's Javelin?

a front-drive small car so I did a quick layout showing the possibilities and got the co-operation of the Austin engine designers in the quite detailed layout of a V4 engine of appropriate size. One of its main features was the use of cheap pressed-steel valve rockers which had just been developed in America and used by Chevrolet. I showed the drawing to Jo Graves, one of Lord's more sympathetic henchmen, who arranged for it to be costed. The result was very reasonable and the Austin drawing office reckoned that it would be sufficiently smooth without the need for a balance shaft. It would not have produced the power of a good ohv or twin-cam arrangement but, as a cheap sturdy four-cylinder, it might have well satisfied BMC's requirements. Some years later I was told that the Austin side had also experimented with the narrow V engines based on Lancia designs.

Presumably the Issigonis Mini-Minor design using the A-series engine appealed to the impulsive Lord. The concept was appropriate for the concerns about fuel supplies after the Suez Crisis. Did he also think he could have the best of both worlds, revolutionary headline-grabbing features without the tooling costs of a new power-unit? Looking back there were too many untried features at the same time and the failure to work out the financial implications swallowed up far more money than would have been employed to tool up for the V4. Once Issigonis's return had been secured my position as chief engineer was a problem for Lord. He did not want the renowned designer reporting to me, and Issigonis himself did not wish to share responsibilities.

THE LONG-LASTING WOLSELEY 6/90
THE AUSTRALIAN EXPERIENCE 1955-1998

Because of the relative lack of corrosion many cars have survived. Fred Holmes writes from Montrose in Victoria: 'Two Wolseleys that have left lasting impressions are the 6/90 Series I and Series III (with leaf springs) both of which we still own. From the start of a trip they display a sound, solid feel and plush comfort. One can feel the advantage of building the car on a very strong chassis. Road roughness does not faze this type of design as it takes on these surfaces with no rattles or squeaks and with no vibrations through the steering wheel. The Series III, in particular, with its timber dashboard, right-hand gear change and many refinements, is still a very pleasurable, reliable classic that an owner can proudly drive anywhere with confidence. We do.'

Peter Richardson from Barwon Heads, also in Victoria, writes: 'I first saw a 6/90 at the Melbourne Motor Show in March 1955. Later that year I had to spend a year at Greenwich to further my training in the Royal Australian Navy. On arrival in the UK we went straight to the Nuffield

Exports showrooms in Piccadilly and learnt that, if we were prepared to forgo our original choice of red upholstery, we could have a black export model 6/90 Series I with brown interior later that day. My wife, Joy, went by train to her mother in Plymouth. In the evening I collected RUV 475 from Eustace Watkins and was in Plymouth by 1am the next day after a careful and comfortable journey. I really only pointed the Wolseley in the right direction. In the morning we took photographs of one of the most graceful cars on the road at that time. During our subsequent stay we learnt that the 6/90 was sure-footed in snowy conditions, although the snow did build up on the windscreen.

'Forty-two years and well over 100,000 miles later we still have and use this Wolseley and I still think it looks good alongside any other car. I have always appreciated the road holding with the built-in understeer. Others have criticised the coil-spring rear suspension but I have found it most satisfactory. (The car was recalled to fit the additional tie-rod and welded bracket but there have been no problems with this bracket or the original welding.) Only last year we completed a 1,000 mile tour of Tasmania. I drove on the rough and gravelly forestry roads at quite high speeds and there were no handling problems. In fact we encountered snow for the first time in over forty years but the car gave no trouble. We never even lifted the bonnet. The engine started first time each morning and idles so very smoothly and quietly. There has been some refurbishing over the years. That engine has been carefully rebuilt. First gear became noisy and I have made up a quiet gearbox with some parts, including an overdrive, from other Nuffield cars. Lights and other equipment have been updated. There is a fault in that the boot lid torsion bars have not held the weight of the lid for years, but this is a minor matter. The car has lasted because it was a good design to start with. I'd like to thank and congratulate Mr. Palmer for his work.'

The Holmes and Richardson long-term experience of the car confirms the Autocar's comments in their road test of a 1957 6/90 Series II (leaf rear spring car).

'One of those rare cars which create an immediately favourable impression but which contrive to improve on acquaintance to such an extent that one parts with them at the end of the test with real regrets. It is a very good car indeed ..... It is unusually comfortable and pleasant to drive, has a creditable performance and a higher than average safety factor in terms of braking and road-holding.'

The 6/90 design could not have been that bad eighteen months earlier, certainly not bad enough to warrant a talented designer's dismissal. The car might have been even better if the engineers had persevered with the coil rear suspension after Gerald's departure. With hindsight, was that September 1955 Harriman interview a fateful turning point towards B.M.C.'s decline?

The Wolseley 6/90 bought by Peter and Joy Richardson when they were in England for a year. This photograph was taken in London on Boxing Day 1956, but the car still looks just as good.

Chapter Twelve

General Motors Experiences

I was very depressed for a few weeks following my resignation. The immediate financial anxiety was relieved when I received a pay-off of £7,500 and a Magnette to keep. This was the largest lump sum that had come our way and generous in those days before the golden handshake. I suspect that the impetus for this payment came from Lord. Whatever his failings, this was typical of recorded examples of personal kindness. I was also concerned for the engineers at Cowley who had been my colleagues. Issigonis may have been a better man for the job but he was not the best of managers. There was too much of what he wanted rather than his listening to other ideas and suggestions. The conventional rear drive cars which followed were competent, strong and reliable vehicles but they did not have the suspension developments which I believed would be necessary to compete with more advanced European designs. Because of inter-leaf friction, leaf rear springs can not compete with coils when the best possible ride is required.

I did have friends and acquaintances in the industry. Max Goldsmith invited me to the Paris Show in October, the first display of the Citroen DS which emphasised the need for better suspension. Others, such as the aforementioned Hennessy, and Lyons of Jaguar, invited me to interviews. However, it was a pre-war acquaintance who approached me with a proposition nearest to that which would suit me. This was Maurice Platt, chief engineer of Vauxhall Motors (part of General Motors), whom I had met before the war through our association with the I.A.E. He had been a motor journalist of considerable note and had left that profession to join Vauxhall as an assistant engineer in 1937. He was promoted to chief engineer in 1953 after Harold Drew, the previous chief who had been at Luton since 1927, had moved to New York as Chief Engineer of General Motors Overseas Operations. The position Maurice offered me was that of Assistant Chief Engineer - Passenger Cars, to be confirmed by the managing director, Philip Copelin, and the Vice-President of General Motors Overseas, Ed Riley - recipient of the American Navy Cross in the First World War and an energetic advocate of the overseas expansion of General Motors. I met both these gentlemen in company with Maurice Platt in the London office. A particular memory is of Riley's interest

Auto-Architect

in Gottlieb Daimler's work in the 1880s and we discussed how Otto's ignition had been improved to achieve the higher speeds needed for vehicle use. There was scant reference to the job but, much to my relief, the position was confirmed. The status may have been lower and I was not a director, but I was delighted to join Maurice's team and relieved for the family finances. I joined General Motors in January 1956 with a considerable increase in salary compared with any I had previously received.

After our previous exile to Ilkley, this time we decided to stay amongst our friends at Iffley. Our daughter was settled at Greycoats School and we were enjoying Orchard House. I soon worked out a route avoiding Aylesbury and major roads (Thame, Haddenham, Stoke Mandeville, Ivinghoe, Whipsnade, Markgate). I left at 7.20am to be in my office by 8.30am and the hour drive became a useful test run for Victors, Vivas and other competitive products. Of the latter, the Volvo P1800 Coupé (assembled in England by Jensen) and an early Rolls Royce Shadow, the Grylls design with self-levelling suspension, stick in the mind. The Volvo was a solid machine but the unresponsive handling did not appeal, and that early Shadow was just too floppy, not taut enough for the cross-country

Vauxhall at Luton. Maurice Platt, centre, was the technical editor of *The Motor* magazine pre-war and Chief Engineer of Vauxhall Motors when this picture was taken. Standing: David Jones – body stylist, John Alden – Commercial Vehicle Engineer, Gerald Palmer – Passenger Vehicle Engineer, Tom Stott – Transmission Engineer, D. Perkins – Administrator, Engineering Department.

110

journey. I particularly enjoyed the Fiat 125 saloon with its zippy twin-cam engine, impressive roadholding, and then uncommon five-speed gearbox. In sixteen years I did not run off the road (I'm not an aggressive driver) and the only incident was when I had to turn back on Ivinghoe Beacon in 1962, the road blocked by snow which would have even defeated the front drive Minor.

On taking up my appointment in the engineering block at Luton I could not but be struck by the influence styling had upon Detroit inspired designs. In fact, the styling activity was akin to a separate department within the engineering department; a holy of holies, a no-go area to any but the privileged. At the time I speak of, the restrained styling of the pre-war models was about to be replaced by the extravagant, almost vulgar, styling coming from the Detroit studios of Harley Earl. I refer to the Victor and Cresta with wrap-around windscreens. Drafting practice was very much like that in the UK, although much of the work on the Victor was done in Detroit entailing a major liaison job with the Luton office.

The other noticeable feature about the department was its close liaison with the accounts department. A senior staff member of the latter was resident in the engineering department and was an important member of the finance committee dealing with new models, or changes to existing models, thus enabling instant cost estimates to be made. These were of great help to the design team. It was the policy at General Motors at that time for each of its makes to have a newly-styled body every three years, with face lift changes every intervening year, so there was always plenty of work going through the drafting offices. In this policy they recognised that the automobile industry is as much a fashion as an engineering business and were rewarded with great success.

Diana shakes hands with Harlow (Red) Curtice, the President of General Motors, during a reception at the Savoy Hotel, London.

Auto-Architect

In March 1956 I went on my inaugural pilgrimage to the States accompanied by my sponsor, the gregarious and charming Maurice Platt. It was a memorable time for me, being my first transatlantic crossing. We flew in one of the BOAC fleet of six Boeing Stratocruisers. These had been ordered when there was no British equivalent available for a non-stop service. They were developed from the Boeing Superfortress bomber and had one of the first civilian pressurised cabins, one actually went upstairs to bed. Our first destination was New York where we attended Motorama, that annual extravaganza of offerings from General Motors for the coming year. Here I was able to meet many of the company top brass, including Harlow (Red) Curtice, who had succeeded Charles Wilson as president in 1954, and Charlie Chayne, vice president of engineering. I must admit I was somewhat taken aback by the very free use of Christian names in most of these introductions. But when it came to talking to the Vice President of Engineering (who was God Almighty to me!) I faltered and stiffly called him Mister Chayne! I had a curious feeling he appreciated the status and dignity I accorded him and we got on well together. This air of familiarity was genuine enough in most cases and often led to much appreciated invitations and domestic hospitality. It is no fairy tale that the Americans as a nation are a friendly people – it certainly proved to be so in my brief visits when I was entertained in different homes and at outdoor events.

On my second trip, again with Maurice but also accompanied by David Jones, Vauxhall's talented styling manager, we were all invited to Ed Riley's farm in Bucks County, Pennsylvania. Shortly before leaving home, I had spied an advertisement for a Mercedes racing car in *The Autocar* and had found it to be the 2-litre machine which had won the Targa Florio in 1924. Subsequently brought to England, the Mercedes was driven by Raymond Mays at Shelsey Walsh and then raced by Lord Ridley until the engine blew up on Southport Sands. It was being sold by Arthur Jeddere-Fisher, well known in vintage car circles, who was about to return to the Fiji islands as Chief Magistrate. He was thus unable to continue

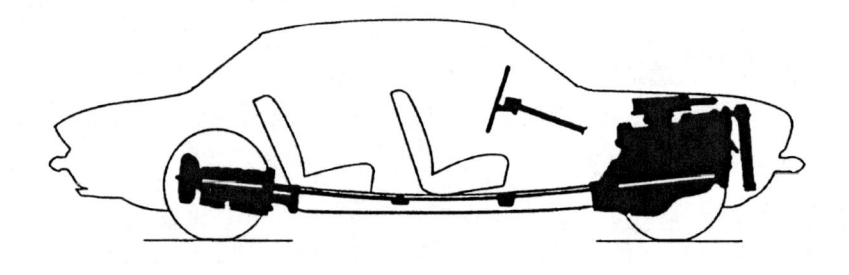

The novel Pontiac 'rope-drive' transmission

the very considerable restoration job needed to save it from the graveyard. I was, so I acquired it for a modest sum and spent the next ten years making it work. *En passant*, Arthur told me that there was another version of this car which competed at Indianapolis in 1923 and that it was in the museum of Henry Austin Clark at Southampton, Long Island. With the prospect of examining this car on the one free day in New York, David and I apologised to Ed Riley for declining his kind invitation. He was very understanding but one or two colleagues raised their eyebrows at the audacity of a raw recruit turning down an invitation from the General himself!

There were no recriminations. David and I went to Southampton, met Clark in his home-cum-museum and he wheeled out the Mercedes. It was indeed an Indianapolis car and I spent several hours making notes and taking photos, all of which helped me in the restoration of my recent purchase. I found Henry a colourful character whose family had made a fortune from Cuban sugar before the days of Fidel Castro.

Of the outdoor events, the most memorable was when I was included a few years later in a small party flying by private plane to see the 1965 Indianapolis 500 mile race. I came away with three impressions: first of the literally hundreds of planes in the private plane park, second of queuing up for a ticket next to of all people, Roy Lunn, who had followed me as chief designer at Jowett and had emigrated to the USA in 1958 to become head of US Ford Advanced Engineering, and third, the race having been won by Jimmy Clark, a Scot, the immediate provision of a Highland pipe band for the victory parade - very slick I thought!

The main point of my first visit with Maurice in 1956 was to examine and monitor the results of the 100,000 mile endurance test, at General Motors's Milford proving ground, of the forthcoming Victor which was an important model in Vauxhall's programme. Milford had been established in 1924 after searching throughout Michigan to find a suitable site. By the 1960s it had grown to over 4,000 acres with facilities for indoor tests and garages spread out so that Chevrolet, for instance, could test separately from Buick. There was a clubhouse where visiting engineers could sleep and eat. It was the pioneering effort for the synthetic testing of motor vehicles and has been copied in every motor vehicle producing country in the world. To see this vast complex of testing equipment for the first time was a mind boggling experience and made one realise the paucity of British equipment and methods. I reflected on what might have been if a fraction of the resources had been available for the testing and development of the Javelin. One felt that a vehicle which passed these stringent tests should be free of service defects for its life. Of course, targets and expectations are continually rising, so that it is not always or often the case. Always on these massive test programmes a sprinkling of suitable competitive cars was included so that a comparison with competition could be made. In fact, the Victor did come through this gruelling test satisfactorily and any defects were notified to the Luton plant in time for corrections to be made before production started. Mechanically this much maligned model was very good - it was its appearance that did not appeal to the British buyer.

Our mission accomplished, we spent a few more days in Detroit for the purpose of getting up-to-date on various new projects, mainly at the General Motors Technical Centre. These included a fuel injection system, and an advanced triple-turbine transmission, but for my part I wanted to see the production of the so-called wrap-around windscreen as we were to use this feature on the new Victor. Therefore, a visit was hastily set up with the Libby Owens Ford Company at Toledo where I spent a whole day seeing the entire process from glass making to sag-bending on frames and zone tempering. It gave me valuable background knowledge in dealing with problems arising from windscreens from our own domestic supplier, not that these were many. The main need was to maintain a consistent body aperture into which the windscreen fitted. In my opinion it was an over-rated sales and styling feature and I was glad it was dropped on subsequent models.

Whether it was intentional or not, routine visits to Detroit often seemed to coincide with exploratory General Motors projects which were probing future design trends, and upon which a wider global opinion, such as that of overseas engineers and market men, was being sought. Often one was asked: 'Would you like to produce this in your territory?' Projects which came into my ken, and which I was shown, were front-wheel drive, rear engines, the flat floor (by eliminating transmission bulge and drive tunnel), and the aluminium cylinder block.

The front-wheel drive project came from Oldsmobile who had designed a front-wheel drive conversion of their V8 engine and attached it to the front of a two-door hardtop. The Toronado was a massive car, and rather crude, and was of no interest to Vauxhall as it would never have sold on the British market. Two other projects were more in line with our size of car, although we could not enthuse over either one. There was a move in General Motors at that time to restore true three-passenger accommodation to the front compartment, which had been reduced particularly on the new compact cars by the smaller seats and by the ever-larger bulge on the floor to house automatic transmissions. Two lines of attack were being considered, and were well under way when Maurice Platt and I were shown them on another visit to Detroit. The first, strongly advocated and directed by Ed Cole, chief engineer of Chevrolet, consisted of transferring the whole of the power and transmission units to the rear of the car. This was a layout similar to, and made familiar by, the German Volkswagen. Due to the larger and heavier six-passenger body a more powerful engine was required, so a six-cylinder, flat, air-cooled unit had been specially designed. It was in an advanced stage of development when we examined it and shortly afterwards it went into production as the Chevrolet Corvair. There was a good deal of controversy at the time on the doubtful handling characteristics of the VW. Because of the similarity Vauxhall declined to take on the Corvair, a wise decision in view of the subsequent activity of Ralph Nader in denouncing this model for safety reasons.

The second, sponsored by Pontiac, was invented by their brilliant young engineer John Z. Delorean and became known as the Rope Drive. The conventional rigid tubular propeller shaft was replaced by a solid 0.625 diameter steel shaft which transmitted drive from the engine flywheel or clutch to a rear mounted transaxle unit. Between these points the shaft did not follow a straight line, but was forcibly held in a downward direction to the shape of a catenary, the lowest point of which was amidships at floor level, the curve rising forward to the engine and rearward to the transmission. The floor tunnel enclosing the shaft was, therefore, much smaller than for a normal drive line, being zero amidships rising to about half its normal height at front and rear with no obstructive bulge for the transmission.

We met and had a long conversation with Delorean about his novel arrangement. He assured us that extensive fatigue tests had been carried out on the bent shaft to arrive at the best configuration and the best material. This project was also in an advanced state of development when we examined it, and shortly afterwards went into production as the Pontiac Tempest. Although it achieved the objective of better front compartment accommodation and a flat rear floor, it involved too many items of unknown service history. Furthermore Vauxhall production was mainly of four-passenger cars in which the advantages of the curved propeller shaft were not so great. In fact, I was later told that all compact sales had been disappointing. This was a consequence of their subsequent poor reception on the strong secondhand market in the deep South where the big families preferred the older designs with much more room in the front.

After advancing rapidly in the engineering/management hierarchy of General Motors, Delorean left the corporation to launch a controversial car bearing his own name in Northern Ireland, with the help of British Government funds. I remember discussing this project when visiting Harold Drew and his American wife in the house to which they had retired at Walton-on-Thames after their years with General Motors in New York, and later in Detroit. We were sitting in their lovely garden and the ever-wise Harold turned to me: 'Gerry, you know its a recipe for disaster'. And so it turned out.

The last of these projects was the move to introduce an aluminium cylinder block to reduce the weight of the small V8 engine which powered Buick's new compact model. Up to that time aluminium had always been regarded as a rare and expensive metal in Detroit and had been little used in major engine components. To counteract the increased price, the General Motors Technical Centre in collaboration with Buick had developed a new process in which the molten aluminium was transferred from the smelter to an immediately adjacent foundry in heavily insulated containers. Money was saved by not having to re-heat the liquid metal. Cast-iron cylinder liners were developed by the Technical Centre. These were cast into the block and had a rough unmachined external diameter to key into the aluminium.

For two years in the early 60s these engines were used in the Buick Skylark and Oldsmobile F85 Jetfire. However, buyers still preferred the bigger conventional cast-iron V-8s ('no substitute for litres' when fuel saving was not an issue) and General Motors could not justify the extra expense at that time. The American market's loss was Britain's gain. Rover directors noticed an example of the engine when visiting America. Bernard Jackman (whom I had got to know when he managed Lockheed and who had returned to Rover as director of manufacture) realised that I had visited Detroit. He telephoned to ask why Buick were giving up the unit. I was able to assure Bernard that it was technically excellent and that its demise was for financial and marketing reasons, neither of which would apply for a more expensive vehicle and in England. I have always been rather pleased that Rover successfully used the aluminium V8 in successive saloons and a development is still used in Land Rovers today.

Chapter Thirteen

Victor and Viva

The Victor with the dog-leg panoramic windscreen had been developed before my time and mainly in the United States. The Fisher Body Division of General Motors had taken over the original Vauxhall ideas so that Luton could concentrate on the six-cylinder Velox and Cresta design. The F-type Victor shape was not admired in Europe but the car confirmed that there was a place for Vauxhall in the 1.5-litre market. I was responsible for a British team charged with reclothing the proven mechanical components with a style more appealing to potential buyers.

Unlike BMC, there was no wish to glorify an individual designer. We all worked together reporting to Maurice Platt; David Jones (who had set up the styling studio in 1937 after studying at the Royal College of Art) worked with Tony Cooke on the body side, Tom Stott as transmission engineer and Arthur Larking as power unit engineer. David created a restrained and attractive shape for this new FB Victor. He had been too inclined, for instance with the Cresta, to dance to the American tune.

My specific design task was to re-work the front suspension mounting. The first Victor had the suspension mounted on the typical American-style detachable front cross member. Colloquially known as the 'bath tub', this seemed unnecessarily large and heavy in contrast to the lighter sub-frame of my comparable 4/44 and Magnette designs. I calculated that the suspension loads could be fed into the integral body-chassis structure by welding large J-shaped pressings, looking like hockey sticks, either side of the wing valance and scuttle assembly. A fabrication at the bottom of these pressings also carried the engine mountings and only a light and simple cross-member was then needed to connect the two sides.

The FB Victor was well received but we were increasingly aware of the limitations of the leaf spring rear suspension. We had developed the higher powered VX derivative and the *Autocar* test reported: 'The suspension feels to be a compromise, one can never really forget the large unsprung mass of the back axle which patters about on some surfaces.' Initiated by the Opel engineers, the

Auto-Architect

subsequent FD and FE Victors had coil springs, trailing arms and a panhard rod. There was some satisfaction in their decision to develop a rear suspension related to my design for the Pathfinder fifteen years before.

The history of General Motors' European investment between the wars is instructive. It is said that the first investment, the purchase of Vauxhall for $2,575,291 in 1925, (then a small company producing only 1,500 quality vehicles a year) was made to test the viability of manufacturing cars in Europe. Having established this, the second investment, the purchase of Opel in Germany for $33,362,000 in 1931, was in a larger company manufacturing medium to low priced cars which aimed to expand that market sector and compete with the likes of Ford. This far sighted strategy paid off. By 1938 Opel was biggest in Europe producing in that year 115,000 cars and 24,000 trucks, and in total far outstripping the Vauxhall production of 32,224 cars and 27,474 trucks in that same year. As Opel had much larger car sales, and was less dependent on trucks, their engineering department was biased towards passenger cars. They were also less influenced by American opinion as exercised by General Motors Overseas. At Luton there were strong sales of the Bedford truck introduced in 1931 and the Vauxhall engineering department had greater bias to the truck and coach side.

Vauxhall introduced the HA Viva (left of picture) in 1963, initially with a brand-new 1,057cc engine. The attractive FB range of Victors had been launched in 1961 and were stylistically much more restrained than the previous model. The FB Victors initially had 1,508cc engines and were available in saloon and estate versions as seen here. The VX 4/90 performance model had 44% more power, disc brakes, and improved interior trim, including some mock wood! In 1963 the engine capacity increased to 1,594cc.

Auto-Architect

Whilst one might have expected some co-operation or exchange of ideas between Opel and Vauxhall during the development of the Victor, it was not until both companies realised the need for a smaller car to extend their market coverage that a joint programme began to be considered. We both investigated different designs; front-wheel drive with both fore and aft and transverse engine lay-outs, rear mounted engines, and independent rear suspension. Opel also produced, and were in favour of, a well-engineered conventional machine with some new concepts in the front and rear suspension systems. I took part in the discussions with Philip Copelin, David Jones, and Maurice. We concluded that the time for rivalry was past and much would be gained by producing a British version of this new Kadett. We had already determined that any new project would be compromised if we just used Victor components in a smaller package. So at Luton we adopted the complete Kadett design, together with mechanical components. The only differences were to accommodate local plant conditions and local suppliers, with minor styling changes to suit the British market. Our version became the Viva HA.

I particularly noted the Kadett's rear suspension development. Instead of clamping the axle directly to the leaf springs, the German engineers had allowed for a small amount of independent movement by pivoting a short axle bracket

The handsome FB Victor in De luxe trim, with two-tone paintwork.

Auto-Architect

on top of the spring. In place of a basic Hotchkiss drive, a short torque tube, itself attached to a boxed cross-member, was meant to absorb all the twisting forces of power, cornering, and braking. When I examined the design on a visit to the engineering department at Russellsheim, I thought back to the problems with the Magnette's torque arm and asked the Opel engineers if they had experienced axle tramp under test conditions. They assured me that the suspension performance had been satisfactory, but subsequently we were to discover limitations. The *Autocar's* brief impression of an early Kadett had been circumspect 'On uneven surfaces, rear end rather lively. Driven with verve, rear end slid out rather too easily'. A year later, when our Viva equivalent had been fully tested, the same magazine reported on rear axle behaviour during hard breaking 'axle wind-up causes high frequency tramp which reduces rear wheel adhesion and hence directional stability'. The next Kadett and the Viva HB had a more sophisticated rear set up with coil springs and trailing arms. It was realised that a panhard rod was not necessary on a small car and that lateral stability could be adequately controlled by diagonal upper arms trailing from the frame and attached to the top of the differential casing. The longitudinal lower arms were similar to those used for the later Victor.

By the launch of this HB in 1966, General Motors passenger car design had virtually ended at Vauxhall and would in future be concentrated at Opel, leaving Luton to continue with commercial vehicle engineering. There had also

The huge General Motors Vauxhall plant at Luton covered some 320 acres. The original Vauxhall factory was built on a 3 acre rural site in 1905; the position of that building is edged in white in the right foreground of the picture.

Auto-Architect

been managerial changes. Maurice Platt had retired and his place had been taken by John Alden, my capable and energetic opposite number on trucks and coaches. However, despite the changes there was still a place for me and I took over responsibility for all power units.

If the day-to-day work was not as interesting as original design, I still had a salary which allowed other activities. The Mercedes restoration (see Chapter 15) was under way. We also had Majorca. When I was working at Cowley, we had continued the holiday journeying which had started with the Javelin as a break from the rigours of Yorkshire. We had explored Spain and got to know some of the Nuffield Agents. Andres Darder (who had been pushing for Minors to be built at Pamplona) had invited us to stay at the new hotel which he had built at Cala de Sant Vincenc on Majorca's north coast. After several more visits, and buttressed by the security of my position at Vauxhall, in 1961 we bought a small single storey cottage which had been built for a local family before the Civil War. I know I am not an architect of houses but I remain quite pleased with the arches I designed for a new enclosed terrace. This was built by one of the Ruisech family, the local builders. My later addition of the tower bedroom was not so universally admired but it served its purpose. In the subsequent Luton years, it always gave the greatest pleasure to be swimming in the Mediterranean on Sunday evening, then to be awoken soon after 3.00am for the 60km taxi ride to the airport. If all went well, the B.E.A. Viscount landed at Heathrow by 7.30am and I arranged for a car to meet me so that I was in the office by 8.30am. We were able to entertain many friends in Majorca, amongst them George Wansbrough's always appreciative widow, Nancy.

The Viva-style HA Bedford van appeared in 1964 and is seen here with the CA van in the centre of the picture and a TJ chassis with a 35cwt body on the left.

Back at Luton I had to keep an eye on all engine development. There were not many problems with the new overhead-camshaft Opel engine but carburation and other settings had to be organised for the full Victor range. Carburettor type, power output and torque were different, for example, between the manual car and the automatic, and different again for the VX 4/90. Then, to fit the Cresta engine into the Victor shell for the Ventora, there needed to be changes to the accelerator linkage and to cable and pipe runs. There was also a different sump to clear the front cross-member. The major design challenge, in which I was supervising the excellent work of Reg Pauley and others, was to create a new 330cu ins six-cylinder diesel for the truck range. There was a particular problem with all our lorry engines when I was in the hot seat. We were always seeking ways of reducing wear in the cylinder bores. One of our engineers, Arthur Larkin, had been working with Laystall to use their chrome slip-fit liners. Initial trials had been trouble free. It looked as if there would be great advantages, justifying the name Chromard. Unfortunately, it was not to be.

When we were in full production, with fleets sold and in service with valued customers, like Danish Bacon and Sainsburys, I began to receive worrying reports of sudden engine seizures. Returned units were stripped for investigation, and we soon found the cause. In order to prevent slippage, these liners had a small flange at the combustion chamber end of the bore. Gases from combustion were

Our cottage on Majorca was a single storey building when we bought it but we added a tower bedroom and enclosed terrace, a chance for me to try my hand at being an architect!

getting behind the liner and causing a carbon build-up between the liner and the cylinder bore which, after much too low a mileage in service, was distorting the liner to such an extent as to reduce the piston clearance and cause seizure. There was no cure but to dispense with the liners and revert to plain cylinder bores, which meant supplying a new cylinder block in each case. This was very expensive. Not only did it cost the company dearly, it also did not do the Bedford name any good. I found myself visiting fleet managers with the service manager, often accompanied by John Alden the chief engineer, to explain our problems. Some users remained loyal but others turned to competitive vehicles.

For the last four years before retirement I was responsible for safety and quality control. This meant keeping up to date with new laws in every market and making sure that the Vauxhall products complied. If it was at times boring after creative work, I was at least gainfully employed and had the opportunity to meet others in the industry. It was a window to the world of the bureaucrats whose usually well-meaning directives and regulations hassled the engineering departments. I also attended the Department of Transport for discussions with civil servants about crash barrier tests.

I was a member of the S.M.M.T. Metrication Committee where we talked about the change over from SAE unified to metric threads, and one became a witness to some of the difficulties in other manufacturing firms. So far as Vauxhall were concerned, the policy on metrication was fixed when Opel became the lead company for passenger cars. Opel had always used the metric system so it became logical and essential for Vauxhall to follow suit. I believe this decision influenced other manufacturers, particularly component suppliers like Lucas, and

A display in July 1996 of cars I designed, Javelin, Wolseley 4/44, MG ZA Magnette, Riley Pathfinder, and Wolseley 6/90. (Photo: David Johnson)

thus accelerated the process of change. Then there was liaison on subjects like pedestrian accidents with the helpful scientists at the Crowthorne Road Research Laboratories. It was all necessary work in the attempt to share common problems within the industry, but I was not sorry when the time came for retirement in 1972. I had been looking forward to having my freedom.

I did not lose contact with my earlier designs. I was asked to become President of the Jowett Car Club and have enjoyed visits to their meetings, and to the meetings of the M.G. Car Club Magnette Register. It has given me a great deal of personal satisfaction to talk with today's enthusiastic owners. However, I have always remembered a comment made by Dickie Downs, then head of Ricardo, when we were visiting an exhibition at the Design Centre in the 1970s. I mentioned that I had recently met the restorer of the Ricardo-designed engine of a Tourist Trophy Vauxhall. Dickie at once responded: 'That's interesting but better still if he had been designing an up to date version'. As I well know from my own efforts with the Mercedes, this nostalgia business needs to be kept in proportion!

Chapter Fourteen

Bugatti and the Work with F.J. Payne

Throughout my sixteen years at Vauxhall, I never lost my great interest in racing and competition cars in which the most advanced of the various relevant sciences and techniques came together. These machines are also generally aesthetically satisfying, which is usually the case when products are designed with the utmost efficiency and simplicity of purpose in mind, and lessons learnt on the racetrack are often incorporated in future road-going vehicles. For complex reasons – financial, political, temperamental – it was the well-known but smaller makes, such as Alfa Romeo, Bugatti, Delage, Mercedes, and Bentley, that were prominent in racing during the inter-war years.

Having lost the opportunity of working directly in this field while I was with M.G. (although there was some involvement through the twin-cam engine), I always had the ambition to own a car of this period and use it mildly as a hobby, so long as the expenditure was within my modest means. I was extremely lucky as I was able to acquire a Bugatti and a Mercedes and their restoration gave me great interest and pleasure. In this work I was assisted by John Payne whose company F.J. Payne Ltd. at Eynsham (started by his father in Oxford in 1905) had established an enviable reputation for high class light engineering skills, which included engine restoration.

From early years I had admired that jewel of a car the T.35 Bugatti, truly called *Le pur sang* and had yearned to own one. However, better judgement prevailed and I chose to restore a T.43 Bugatti as being more practical with four seats and often fitted with the attractive but longer Gran Sport body. I soon found that restorable T.43s did not grow on trees but had to be discovered. Here again, Dame Fortune smiled on me in the person of a vintage car dealer who was advertising, just at the right time, a T.44 rolling chassis complete with all mechanical units but without a body. I went to see this collection of components and decided that, by cutting and shutting the chassis to T.43 wheelbase and making a reproduction body to suit, I could become the owner of a very delectable looking car which at first glance looked like a T.43 Bugatti, although lacking the distinctive light alloy wheels. So I bought the emaciated T.44 for £1,000. A feature also missing was the

Auto-Architect

screech of the supercharger for the T.44 had a 3-litre, unsupercharged engine and was comparatively quiet. The body was a very simple open affair with a folding canvas hood. I fabricated a steel frame which was welded by Payne, and panelled for me by a local coachbuilder.

I was in the Eynsham works one day observing (and chasing!) progress on my components when I saw Payne himself machining a large piece of steel bar which was patently not a Bugatti part. On enquiring what it was, he told me it was

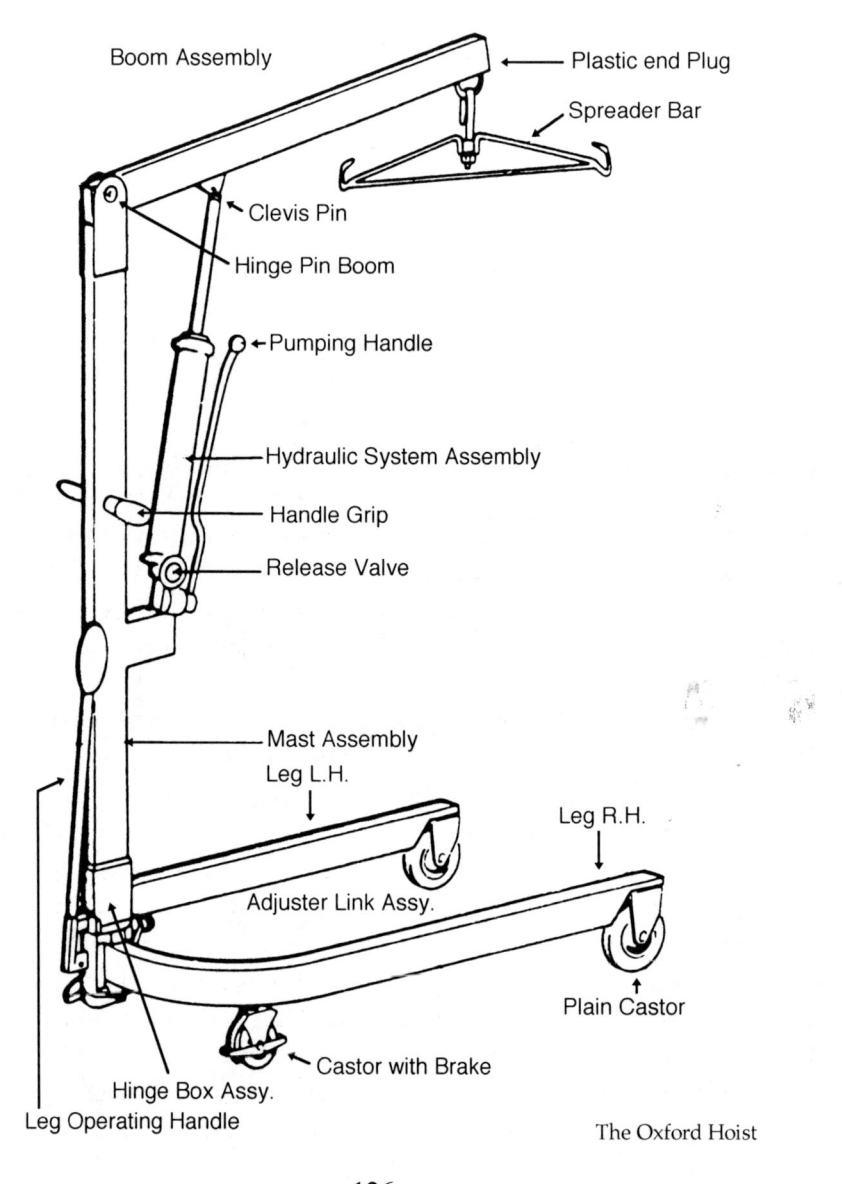

The Oxford Hoist

the body of an hydraulic pump he was designing to operate a small fabricated mobile hoist. This was for lifting a disabled person from wheelchair to bath, bed, or car and thus providing additional mobility. It was being developed at the suggestion of a well known Oxford consultant, a Dr. Ritchie Russell, and one of its main features was a variable rate of descent operated by a sensitive hand control. While appreciating its good points, I thought it was crude and out of date to make the body from steel bar. It would be better made from a casting or forging, with the valve passages re-arranged to minimise the machining. I drew a sketch which I sent to John Payne and which he ultimately accepted. This change made the the hydraulic pump, and the entire hoist, a better proposition for production.

With the body manufacture made easier and a little encouraging market research, the Oxford Hoist did appear to be a viable product and the decision was made to go ahead with it. The problem was where to make it. There was not enough capacity at Eynsham for a new manufacturing section and the only answer was to separate this from his other work and move into a separate building. A new company was formed to control the functions of building the Oxford Hoist. As I had done a fair amount of work on the project in one way and another, John Payne at this stage asked me if I would like to subscribe to the capital of the company and join it as a director with 10% of the equity. This I was more than ready to do as I could see a bright future in any device which helped to make the lives of disabled

The T. 44 Bugatti I purchased and restored fullfilled a long-held ambition to own an example of this famous marque.

people more bearable. The factory was built and production started early in 1976. The new hoist was an immediate success and was soon selling all over the world. In Sweden, a country in the forefront of aids for the disabled, they no longer spoke of a patient hoist but merely of an 'Oxford', so popular had it become.

With this success, came a request for additional products from the Eynsham factory in which I was partly involved. Over the years 1976 to 1996 a smaller version of the hoist was added for domestic use which could also be used at a swimming pool or in conjunction with a car. With the increased use of commodes and shower cubicles at home a special mobile chair was developed with superior anti-corrosive qualities. For bathroom use, we devised a special hoist which enabled a patient to transfer themselves from wheelchair to bath and back with no outside assistance. This was a period of great inventiveness and development in the field and I found it very stimulating to walk around the annual NADEX exhibition, which was devoted to aids for the disabled, to observe the latest ideas from companies great and small, UK based and foreign. It was inevitable that amalgamations should take place, in fact it was to a degree desirable. When John Payne (who had a succession problem) announced he was thinking of selling the company in 1996 I positively supported him as I could see it would bring in fresh capital which was needed to develop other new products. Furthermore, my investment in the company in 1976 had been very profitable, and at the age of 85 it was time I finally retired.

Chapter Fifteen

Targa Florio Mercedes

'BUT M'SIEU, WHERE IS THE REGISTRATION BOOK? HOW DO I KNOW THIS CAR IS YOURS? HOW DO I KNOW YOU WILL NOT SELL IT IN FRANCE?' My heart sank - this was the only document I had forgotten and the customs officer on the quay at Le Havre was threatening to upset all our plans and send me back to England. I was on my way to fulfil a long-held ambition of mine to visit Sicily for the famous Targa Florio motor race and, furthermore, I was taking with me the fabulous Mercedes racing car which had won the race in 1924. It was now 1974, the 50th anniversary of its win and the organisers, the AC di Palermo, had invited me to start the race by driving the first lap of 45 miles of this tortuous Circuito di Madonie. I had read about and followed this event since I was a schoolboy, and could not believe that I was to drive the circuit at last, even in an old timer. So I was determined not to let a minor French customs official stop me from making the trip.

In my best schoolboy French, I pleaded with him, I harangued him, I showed him the papers about the event to which I was going in Sicily, all to no avail. All he wanted was the registration book. Then I suddenly thought 'Why not let my wife deal with him – she is charming and speaks good French?' So I persuaded her to come into his office. In ten minutes of negotiation a compromise was reached – I was to promise faithfully that on our return journey I was to report to him and show him the car. Never have I agreed to a bargain more quickly – HURRAH, AVANTI, we could be on our way!

The party comprised Diana and John Payne, who had kindly agreed to support me on this epic trip and provide the towing car, another Mercedes of much later vintage (a 200D saloon). The racer was one of a team of four, 2-litre supercharged cars entered for the 1923 Indianapolis 500 mile race. They did not perform well and were returned to Stuttgart where they were modified by Dr. Ferdinand Porsche when he joined Daimler as chief engineer, mainly by almost doubling the size of the supercharger. They were entered for the 15th Targa Florio on 27th April 1924, which they won, with Christian Werner driving the winning car at 41mph.

Auto-Architect

Arthur Jeddere-Fisher had discovered the car languishing after the war in the Shuttleworth Collection at Old Warden, but fitted with an incorrect 1.5-litre engine. Through Laurence Pomeroy and Terry Breen, components of a correct engine had been found but, even after work on the top end at Mercedes, bottom end problems remained. Three connecting rods, their bearings and the roller bearings for the crankshaft were missing. Part of the fascination of the restoration lay in finding out what should have been there and, in addition to my visit to the Austin Clark car in America, I examined the sister car in the Mercedes Museum at Stuttgart. Roller bearings are lubricated by just a squirt of oil. I had to somehow find, or draw and make, these parts or redesign them using plain bearings which would need a new lubrication system. I chose the latter course, and, as with the Bugatti, was greatly helped by friends. John Payne ground the main journals on the existing crankshaft and made odd parts like bushes and fork ends for control rods. Hi Duty Alloys forged new dural connecting rods. Tony Vandervell arranged for these to be machined and fitted with Vanwall plain bearings. The next task was to organise high pressure lubrication and I was helped by the Vauxhall apprentices, who were glad to practice their skills on this and numerous other small but vital components (there was a sensible arrangement by which payments contributed to running costs). We altered a geared oil pump from a Bedford engine and this was then mounted so that it could be gear driven from the crankshaft. Hepworth & Grandage, who had been suppliers for the Jowett Javelin, made new pistons at their Bradford works.

The rebuild of that advanced engine, Paul Daimler's last design, was a co-operative effort from British industry. It is one of the earliest designs with four valves per cylinder operated by twin-overhead camshafts and much copied today. The cylinder block is the usual fixed-head Daimler aircraft design, with steel barrels and sheet metal welded water jackets. The foot-operated supercharger gives a considerable power boost and is used mainly for short bursts of acceleration. The whole restoration took the best part of ten years, but I was rewarded with a fine car with near original performance. The first outing was to Luton Hoo where I.A.E. members had been asked to bring significant machines. I then took part in events at the Prescott Hill Climb and at Silverstone, where I won the first race which I had entered with a little help from the handicappers. The car never faltered but the driver may not have done full justice to its capabilities. It is a delightful driving machine, with a top speed of about 110mph and light and accurate steering.

Then came the Sicily trip and, to save wear and tear and conserve fuel, we towed both ways using an A-frame towing bar. As John Payne could spare only two weeks from the business, we chose the most direct route to Reggio de Calabria, the crossing point for Sicily, resisting the temptation of sightseeing on the way. Outward, it was Paris, Chambery, Mt. Cenis Pass, Turin and Bologna. In Italy coupled vehicles were not allowed through the Autostrada toll booths. We had already decoupled on the Mt. Cenis and it was Diana's idea to start up the Mercedes and drive the car through under its own steam. We then drove on separately and recoupled the vehicle when we were out of sight of the booths. In fact the tolls were then too costly for our budget and we transferred to the

Given my lifelong interest in sports and racing cars, I was fortunately able to buy and restore examples of two of the finest ever built. The T. 44 Bugatti chassis was shortened and fitted with a reproduction T. 43 body.

Auto-Architect

super-strada along the Adriatic Coast to Bari, Taranto and then across to Reggio. Homeward, it was Reggio, Naples, Florence, Spezia, Genoa, Turin and thence Mt. Cenis Pass, and the outward route to Le Havre – a total mileage of about 2,500, including about 400 miles in Sicily. All went well on the journey in both directions, the outstanding recollection being the intense interest the *macchina* created, with small groups gathering round it whenever we stopped. One or two handsome offers to buy it were made, and I'm sure I'd have left Italy many million lire richer than when I entered, had I accepted the highest bid and reneged on my promise made at Le Havre!

After the short crossing to Messina, we drove along the coast road for about 50 miles to our final destination, Cefalu, where the AC de Palermo had reserved accommodation for us. Hardly had we unpacked our bags when

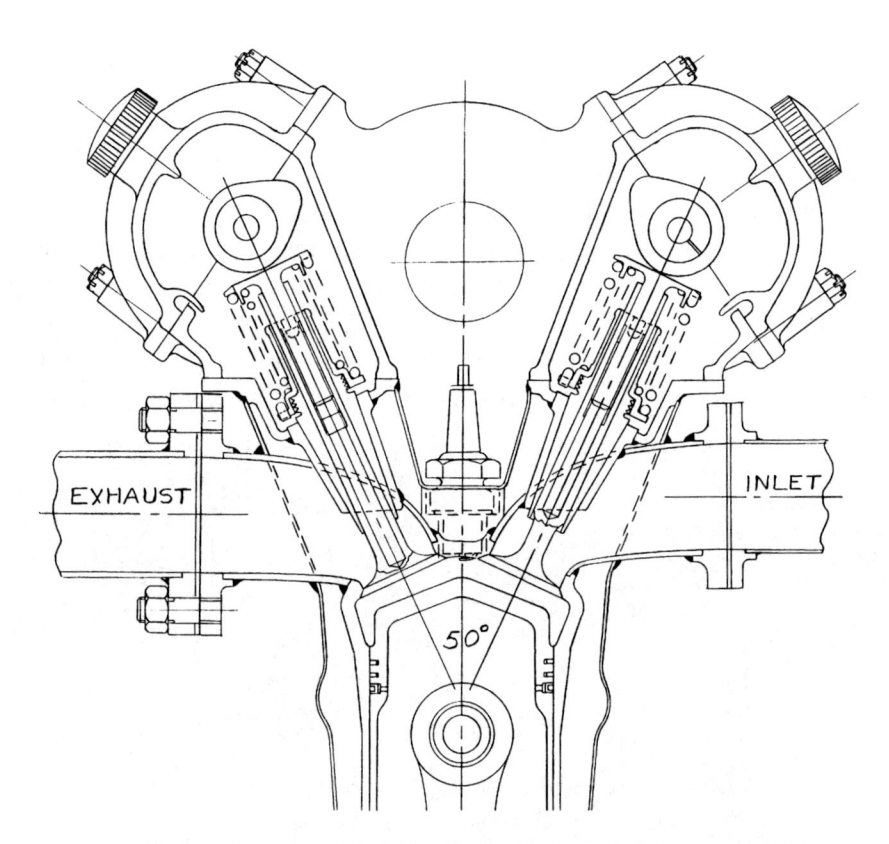

My 2-litre Mercedes won the 1924 Targa Florio and was later driven at Shelsley Walsh by Raymond Mays. As many parts of the original engine were missing when I bought the car, the restoration took me ten years. The engine is of a particularly advanced design, having four valves per cylinder operated by twin overhead camshafts. When working on the car I produced this drawing of the cylinder head. (See also picture on page 132)

an official of the club appeared with an offer to drive us around the circuit to familiarise us with it, for which I was grateful as I had to drive the Mercedes round it the next morning. The Madonie Circuit is on the north coast of Sicily

Once the restoration of the Targa Florio Mercedes was complete, I took the opportunity to enter the car in hill-climbs and circuit races. It performed well, having a 110mph top speed, and at Silverstone, with a little help from the handicappers, I won the first race I entered. I competed in a number of events at Prescott (Above).

(Below) In 1984 we entered the Mercedes in the Prix Maritime and this event gave me the opportunity to photograph my car with the Indianapolis Mercedes owned by Gerhardt Von Raffay.

Auto-Architect

and has been the venue since 1906 for the race for a shield (a targa) presented by a local grandee, Comendatore Vincenzo Florio, who was a keen motor sportsman of the day and who wanted a circuit of demanding roads upon which to test the various cars available to him. He chose one in the wild Madonie mountains near his estates in Sicily and, largely due to his enthusiasm and despite the remote and romantic venue, the event was taken up by the makers of the faster sporting cars of the day and became world renowned. It very soon became the equivalent of the Isle of Man TT in the motor cycle world.

The circuit consists of an eight-mile stretch of the coast road between Palermo and Cefalu, the ends of which are linked by a network of minor mountain roads connecting small villages, thus providing a choice of three circuits according to which mountain roads are used. The lap distances are 92, 67 and 45 miles. They had chosen to use the short circuit for the 1974 race, so I was instructed to present myself and car at the start, or tribunes area, at 6.00am to commence my opening lap, after which the roads would be closed to all but race traffic. This I did, taking with me my wife as riding mechanic in compensation for all the lonely hours she had spent while I was restoring the car. Diana thoroughly enjoyed the experience.

One of the highlights of my motoring life was the chance to cover a number of 'parade' laps of the Monaco Grand Prix Circuit in company with other Mercedes-Benz cars as part of that company's celebration of their centenary. Diana was able to accompany me as 'mechanic'.

There followed one of the most exhilarating drives I have ever made. Left, right, hairpin bend after hairpin bend, up and down from sea level to several thousand feet as the road snaked up the Sicilian mountains. Down to a dry river bed, again up a mountain side, through two small villages, until, in a series of fast bends, it descended to sea level and the long five mile flat-out straight back to the start. I literally threw the Mercedes round this circuit, determined to enjoy every minute of it, and the car responded magnificently. It had taken me about one hour and twenty minutes for the forty five miles, compared with Werner's one hour and thirty-eight minutes for the 67 miles medium circuit in 1924. The Madonie certainly lived up fully to my expectations.

By the time we got back to the start the 50th Targa Florio race had commenced, so that any hope of doing a second lap, which I would like to have done, was out of the question. I also would have liked to have driven round the long circuit of 92 miles, which is much more difficult, but we were pressed for time. After three memorable days in Sicily, and much interest shown by many different people, we reluctantly returned to the mainland at Reggio and started the long drive home via Le Havre, where we kept our tryst with the *Douannier*.

In 1984, Diana and I travelled with Jill and John Sutton (we had met when we were both at Vauxhall) to take part in the Prix Maritime, a series of speed trials on country roads between Hamburg and the Baltic coast. It was organised by Gerhardt Von Raffay who had been the main Volkswagen dealer for Northern Germany and who had then established a museum in Hamburg devoted to German sports and racing cars. Gerhardt had discovered the chassis of another Mercedes Indianapolis racer in California, then fitted with a two-stroke Schmitt engine. Mercedes engine parts had been located in Pittsburg and all these components had been sent for restoration in New Zealand as a result of an encounter with a skilled restorer called McNair, who had worked wonders fixing another car on a previous New Zealand Rally. Our attendance at the Prix Maritime was an opportunity to photograph the two Mercedes together. We used John's Vauxhall Royale to tow the racer to Harwich from where we had a free passage to the German port. In the speed trials John drove and was placed second to a T.35 Bugatti, winning a small trophy. Daimler-Benz then gave us an inscribed silver ashtray as a tribute to the work of restoration with the consequent publicity for the company.

The 1974 expedition to Sicily had been so enjoyable and successful that when I was approached by Erik Johnson, the public relations chief of Daimler-Benz in the UK, to join him in taking part with the Mercedes in a publicity exercise on the occasion of the company's centenary year in 1986 I was 'all ears' as the saying goes. When Erik told me it was to take the form of a parade of well-known racing and sports Mercedes cars around the Monaco circuit just before the Grand

Prix in May, with all expenses, except personal accommodation, paid for by Daimler-Benz, I lost no time in accepting. Unlike the Sicilian trip when we towed the car on its wheels, this time I hired a trailer for the Mercedes which was towed behind a modern saloon supplied by the Company. This gave a comfortable ride for Erik's wife and Diana, who were part of the crew.

We passed through Le Havre where our friend the *Douannier* spied us and this time waved us on as I'd not forgotten the registration book! We travelled on the newly opened Autoroute A13 to Paris, then round the Periphérique and onto the A6. After an overnight stop near Lyon, we arrived at Cannes by late afternoon. We left the Mercedes in the showroom of the local agent with the other participants. These included a 1902 Mercedes and the 300 SLR with which Stirling Moss won the Mille Miglia (both of these from Stuttgart), an SSK from a private collection, and two or three more ordinary models from the vintage period.

The next day was spent in preparing the car for the show and on the morning of the Grand Prix we were escorted in procession to an underground garage that opened onto the Marine Drive, part of the Monaco Circuit. The day's programme included a race for Renault R4 Specials and, as soon as this was concluded, the Mercedes circus was let loose upon the Circuit. As in Sicily, I insisted that Diana accompanied me as mechanic as I realised that this was another red letter day in my motoring experience and that she should join in the fun. Let loose on this historic circuit my exuberance was such that I only moderated my pace when repeatedly flagged by the marshals to slow down. I often think back to that memorable parade on that historic circuit, a fitting conclusion to this tale.

Epilogue

by Christopher Balfour

Revisiting available sources, and without Gerald's veto on any consequent conclusion, there can be further explanation of some of the events in this book. He died in 1999, in his 89th year, and in tribute Javelin and Magnette owners brought their cars to the funeral at Iffley Parish Church. The obituaries praised his talented designs and Anders Clausager, writing in the Guardian (8th July 1999), suggested 'his departure from BMC was conceivably much to its later disadvantage'. If Gerald's thinking on compliant suspension competitive with European offerings and more accurate costing had been accepted he might indeed have been the saviour of BMC. But the personalities who, in 1955, wanted him removed from mainstream design, or feared to contest that directive, would still have rejected his ideas if he had stayed. Sidelined to gas turbine development, he would have had little influence.

It was a clash of attitudes. Gerald studied the available technical knowledge, introduced his own design flair, and wanted to sell, at a profit to the corporation, machines that were at least as good as, if not better than, the products of rival manufacturers. But what he could not do was alter history. At the start of the 20th century the new British motor industry had been largely left to the artisans. Lord Nuffield, that vastly successful early school leaver cycle mechanic, disliked graduate entrants whose classroom learning took the place of workshop experience. In consequence the company was short on cost accountants, economists, marketing and other experts and, on top of this, his Lordship's prickly disposition in later years had led to the departure of competent home grown managers. So, by the time the mercurial Leonard Lord had fought and sworn his way to the top, whilst there were many hard-working technicians answering to his demands, there were fewer widely educated employees with the confidence to question what often seemed to be dictatorial hunches. To stand up to Lord, Gerald would have needed support that was not forthcoming. He may also have realised that only mixed benefit would have come from an improbably successful 'palace revolution'. Whatever the failings, under the Lord regime, reflecting all his experience back to his first job as machine tool engineer at Hotchkiss in 1923, the massively complicated business of bringing men and machines together did produce automobiles. Had an unsuccessful challenge led to dismissal without references, future employment for a colonial engineer without access to the English public school network, and without private capital, would have been much more difficult. In the railway hierarchy known from his childhood, employees did not stand up and fight the boss. Besides Gerald, in the words of one of his colleagues, 'one of the nicest men you could meet - he never says anything unkind about anybody', did not have the appetite for, or experience of, the deceits of power politics. His skills were as an exceptionally talented designer backed by painstaking hard work in order to achieve a desired result.

Current owners and historians both agree that the Javelin was an outstanding achievement. It was the first post-war saloon that handled as well as, if not better than, the so-called sports cars of the time and this virtue was combined with a roomy cabin and a comfortable ride. Except for the decision to switch from solid cast iron to split aluminium cylinder block, a mistake readily admitted by Gerald which affected crankshaft and bearing efficiency already diminished by taxation considerations (see page 45), the car's later problems were not of his making. Had the finance and development capacity been available, the Javelin could have been a world beater. The potential was confirmed by Press comments in the car's last years. In 1953, *The Motor* magazine wrote: 'the purpose was to check the claim that every weakness had been eliminated. Having completed the 2,500 mile journey, including Ghent to Ostend at over 75mph (and no decline in oil pressure during the final mile at full throttle), without a problem, we are pleased to record that the claims for true reliability show every sign of being justified.'

There have been mixed assessments of Gerald's contribution to Nuffield and BMC after his return to Oxford. Some of the commentators seemed not to understand that circumstances, the Chief Engineer appointment in 1952, dismissal in 1955, meant that he did not oversee the development of his later creations. The Magnette and associated Wolseley was the one design just about in production before his promotion where other hands did not alter his concept and harmony of proportion (Except that they replaced the curved front wing chrome with a straight strip after his departure). Still, even today look at a ZA-Magnette and dwell on the subtlety of each line and curve. Nor did the performance disappoint despite the strictures on the use of existing components. Gerald's creation was in every way so much better than the machines that followed and it is a pity that its inbuilt strength meant that in later years so many examples ended their lives on stock car race tracks.

Jim O'Neill has written to me about those times. He started at Pressed Steel in 1936 and, after a post-war spell at Austin, had moved to Morris. He was responsible for the M.G. TD body when Morris engineers were fully occupied

Auto-Architect

with the new Morris and Wolseley designs and in 1949 he joined Gerald's newly established design office as senior body engineer. The brief was at first to concentrate on M.G. This was to have been extended to Riley after Harry Rush's retirement but then Rush was killed hastening this assignment. Alec Issigonis, still then junior to Vic Oak, concentrated on Morris and there seems to have been more uncertainty about future Wolseley design than Gerald remembered in our discussions forty-five years later. Jim writes of understandable friction between Issigonis and Gerald. 'The engine, suspension and body sections were so beautifully detailed in Gerald's quarter scale drawings of the new saloon that it did not take long before the full size layout draft was complete'.

Reg Hanks had included Wolseley in his original commission (see page 67) to Gerald who then realised that a common body shell would suffice. Delay resulted because Issigonis, up until then responsible for Wolseley, had to be consulted and, as Jim continued, he would not give a decision. Was this the start of the difficulties between them; Issigonis already fearing diminished responsibility

Despite various technical problems arising mainly from a lack of sufficient finance to fully develop the cars, the Jowett Javelin proved to be a hit with many drivers. One often hears of praise from past and current owners and one of these, Dr. Clarence Eminson a Lincolnshire opthalmologist, covered an enormous mileage in his Javelin commuting between Doncaster and Scunthorpe and travelling around the country. His journals and diaries record experiences with many cars from the 1920s to the 1970s and he says of this Javelin, purchased at the beginning of 1950 and registered HDT 501: 'I have done 370 miles in ten hours by myself and felt fit enough to keep on indefinitely, which I could say for few other cars I have owned.'

Auto-Architect

and Gerald rather more determined and spirited than later writers suggested? The Magnette version was already decided and Jim remembers Gerald coming to him after a delay of several weeks delay saying: 'I am not waiting any longer. We will raise the car two inches on its suspension and that should give the sit up and beg appearance of a Wolseley saloon'. This is how the two models developed (see chapter eight) and Cowley decided to put the Wolseley 4/44 into production ahead of the M.G.

The sports car proposed to replace the T series (pictured pages 76-79) was another project from Gerald's separate design group. In Jim's view the concept with interchangeable body panels was outstanding. The car with flowing wings that he remembers as being painted pale green was a running mock-up, if not fully engineered. In a 1990s interview talking with Jon Pressnell Gerald had memories of going out in this prototype and also that the strong common base structure clothed in light sheet weighed less than the then current T-series cars. But it was one of many projects that did not see the light of day. In that period before, and indeed after, BMC was formed, the company still seemed to be composed of competing empires, despite Reg Hanks' efforts at centralised control from Cowley. John Thornley was another determined, though different, character to Issigonis and after years of service he was soon to be general manager of M.G.

Jim O'Neill with whom Gerald worked closely on a number of projects.

He preferred the simpler design based around a separate chassis that he and Syd Enever were hatching to Gerald's concept. This became the excellent and popular MGA of which over 100,000 examples were built.

The troubled development of two other of Gerald's designs have cast shadows on an exceptional talent. Gerald conceived the twin camshaft head for the B-series engine and this was originally intended for both Magnette and his interchangeable M.G. sports car. The project was agreed in principal with Chief Engineer Vic Oak in the period before Gerald took over that position from Vic after Issigonis departed for Alvis. The layout was chosen with a view to maximizing reliable power from strong valve gear and efficient combustion chamber shape (see page 75 and a sectioned engine is on display at Gaydon). Gerald was then too busy to oversee the detailed development that was left to others. The subsequent troubles were compounded by BMC central management's orders first that the standard B series block should be used and, second, that when fitted to the MGA (by which time Gerald was with Vauxhall) the car should be on general sale rather than, as Thornley wished, reserved for customers who understood the maintenance and tuning requirements of a high performance engine. That Gerald's original design was sound is confirmed both by the delight of twenty-first century owners in the crisp responsive performance of correctly assembled and tuned engines and the towering performance of EX181, the Roaring Raindrop record car (also see pages 76-77). This beautiful machine, again painted pale green, is also on display at Gaydon.

Gerald's Riley Pathfinder prototype was on the road in May 1952, just a month before he became chief engineer. There is much about production difficulties with this car, and the associated Wolseley 6/90, in chapter nine. Riley production had already been transferred to M.G. at Abingdon so when John Thornley finally became manager in that same year he took overall responsibility for the Pathfinder, ably assisted by works manager, Cecil Cousins. Now that there is access to Thornley's memoranda and other information sources, including research undertaken by David Rowlands (see page 87) and David Knowles, it is easier to understand the car's problems and potential. Once again decisions by others diminished Gerald's genius.

He had wanted to continue the Riley tradition of responsive rack and pinion steering but after the directive to fit the big Austin six (the C-series engine) into the Wolseley stablemate, there was no room for the rack in the common structure. Then, emphasizing the debit side of the central control exercised by Hanks, the order came down from the BMC sales and commercial organisations at Cowley that the Pathfinder must be on show at Earl's Court in 1953 and available to dealers soon afterwards. Thornley protested. He could not guarantee trouble free vehicles. There were still problems, and this was before the visit to John

Auto-Architect

Thompson at Wolverhampton (see page 88). He was overruled, Gerald had no authority over sales and the central directors higher up the ladder stuck rigidly to their decision. The Riley 2.5 litre RMF sales were slowing and they wanted a new product on their stand at the London Motor Show.

It just was not possible to get the car right in time and cope with the Thompson build problems that had not been evident in the carefully constructed prototypes. Early customers loved what they saw, the simple, elegant lines and the superb finish of the show models, but did not like the problems they subsequently encountered. The Wolseley fared better, both because if was not introduced till the 1954 Show and also because there were fewer difficulties with the older style Lockheed brakes without a servo. By 1955 Thornley and Cousins and their team had sorted out the Pathfinder. This was confirmed in the complimentary *Autocar* road test published in 1956: 'long torsion bars at the front and coils at the rear, a suspension suitably firm for high speeds yet comfortable for slow speed driving over indifferent surfaces.... the high overdrive top gives a feeling of being wafted along in comparative silence, the speedometer on the 90mph mark, the rev. counter recording 3,000rpm'.

Thornley believed in the Riley's future as evidenced by his memo to S.V. Smith, then the central BMC engineering co-ordinator, dated November 1955, a month after Gerald's dismissal. Under the heading 'PATHFINDER SHORT-TERM' he said: 'Due to take the twin camshaft C Series (Gerald's idea) when this is developed. Will involve redesign of front end and the opportunity may be taken to make this change fairly radical. Rear suspension should be carefully considered and steering layout deserves a close look. In brief, policy on the Pathfinder, apart from engine change, is to evolve and perfect.'

Alec Issigonis with some of his creations at his retirement party in 1971

And what a car that could have been. It was another ten years before BMW produced their first post-war overhead cam six, the 2500 and 2800. Rear coils were mated to independent rear suspension that Gerald, aware of both African roads and O.D. North's experiments, had been promoting since his student days and had built into the Deroy. By 1957 Cam Gear's 'Hydrosteer' power steering had been fitted to a Wolseley, and in March that year the *Autocar* testers commented on a 6/90 automatic: '...the Wolseley Six-Ninety is a well-designed and soundly built car of real character, which should appeal to the type of owner who appreciates his vehicle enough to keep it for several years.'

I chanced on a 1958 letter from a Riley Pathfinder owner in the correspondence columns: '...a collection of very interesting and well-engineered chassis design points. Just when the difficulties had been put right, it was withdrawn. Heavens knows why. A tireless mover at high speeds with true Riley handling qualities. Last summer on the Milan – Turin autostrada our Pathfinder covered thirty miles in twenty minutes, a constant 90mph, at which speed the engine will cruise all day on a low throttle opening. When leaving at Cavaglia, oil pressure and temperature gauges showed normal readings. I hope BMC will soon return with a similar Riley car.'

By then it was far too late. Gerald had been at Vauxhall for over two years. His advocacy of torsion bars and coils, constant since the 1930s, had been decisively rejected. The Pathfinder had been altered to leaf rear springs. The subsequent 2.6 had the standard C series engine. In 1959, instead of 'similar Riley cars', the corporation introduced the cumbersome Farinas, the 6/99 and A99, backed up by the A55, Oxford and their derivative, the Magnette III. The latter, in particular, was a disappointment for enthusiasts as in all respects it was not a patch on Gerald's appealing ZA. All had nineteenth century 'cart' rear springs and the magazines commented 'too much roll, road rumble, low speed harshness'.

Leonard Lord, firmly enthroned as BMC dictator, believed that Issigonis, even though he was away working at Alvis, was the man for the corporation's future smaller products. He did not want the tooling costs of a new engine (see page 104), he was aware of Jack Daniel's continuing work on the Issigonis ideas, and the personal relationships seemed to work. For all these reasons, when the time came to build a new small car Lord preferred the Mini and Moulton rubber suspension as the concept instead of Gerald's wishes for a properly costed front wheel drive range with modern engines, better use of space, comfortable interiors and well engineered metal suspension. The writing was on the wall when Gerald was ordered to approve Burzi's replacement Minor (see page 105), so much so that Gerald dismissed his own Minor replacement prototype seeing little merit in just another frock clothing an existing design, which had no advances in his areas of concern.

Auto-Architect

Instead of summoning Gerald to discuss Pathfinder realities and future plans, the immovable Lord, who of course did not want his resolute vision clouded by alternatives, latched on to some minor Wolseley criticism (it could have been anything, the road tester should not be blamed) in order to give Issigonis the free hand which, totally in character, he was demanding as the price for his return (see page 109 'too much of what he wanted'). Personal antagonism against Gerald was not part if it. Issigonis had resigned before Vic Oak's retirement and Gerald's elevation. It was more that he had every detail of the Mini in his mind and input from another talented designer would have compromised and delayed the project. The Alvis sabbatical worked well for Issigonis parked away from BMC turmoil till the time was ripe for his 'charwoman's car' (see page 68).

It may be that the Mini is now considered an icon worthy of a BMW copy/enlargement (though without Moulton suspension) and its history is much praised in the enthusiast press. But what did the original design do for BMC? Where now is the Corporation? And why, ten years later, were Audi, Fiat, Peugeot and others producing the designs that Gerald had wanted? The Issigonis Mini replacement, the 9X, is on display at Gaydon. It encompasses all Gerald's thinking, the Moulton suspension replaced by coils at the front and torsion bars at the rear, more passenger space, less weight and more potential profit resulting from lower costs. But it was not ready till the late 1960s. By then Lord had long retired, the British company had lost the small car market to the Europeans and

The Issigonis 9X proposed Mini replacement was closer in concept to Gerald's thinking than the original version. With coil front suspension and torsion bars at the rear it would have been cheaper to build.

the new bosses from Leyland didn't want that sort of car. Issigonis himself was sidelined with Harry Webster from Triumph appointed as technical director reporting to Donald Stokes.

Their rear drive offering was the uninspiring Marina. From our many conversations, and the evidence of Deroy and Javelin, I believe Gerald's thinking was echoed by Gerd Schuster, project leader for BMW's (yes, them again) rear drive one-series in 2004. 'We wanted a high quality of steering and precise handling in a car that does not under or oversteer'. Gerald realised that there would always be drivers who did not like the compromise resulting from using the same pair of wheels for propulsion and steering. He wanted responsive, pleasing rear drive cars as well as the space-efficient front drives, and in Riley and M.G. there were marques that could be thus developed.

There was less opportunity for innovative design at Vauxhall. But what else could he have done? Ford, to whom he had been suggested by Hanks, remained wedded to the effective but orthodox, witness their success with the Cortina. To further his own concepts would have meant working in another country and neither he nor Diana wanted to uproot from Iffley. Other manufacturers did develop these layouts and a rear axle beam doubling as anti-roll bar is common practice with front wheel drive. Gerald also realized that mechanical innovation alone did not sell cars.

Buyers wanted reliability and convenience with fashion increasingly playing its part alongside engineering. Vauxhall offered solid work on conventional designs, interesting contact with General Motors, and a salary that allowed enjoyment of the rest of life. Mindful of the Issigonis star status, Gerald always insisted that individual designers were not glorified at Vauxhall (see page 117). Yet it's not just coincidence that the first Victor after his recruitment, the FB, had notably attractive lines. Gerald gives all credit to David Jones yet it was his responsibility as team leader to sanction, perhaps even encourage or adjust, that particular shape from a designer who 'had been too inclined to dance to the American tune'. We do not know what might have resulted without his influence.

The J-shaped pressings illustrated in the *Autocar* in September 1961 followed on from lessons learnt with the Magnette at BMC and contributed to weight reduction. He still had to accept his old hate, the leaf rear spring; but the Opel engineers shared his thinking and would have been aware of his earlier work when a coil rear suspension related to the Pathfinder design was incorporated in FD and FE Victors.

Because the design committee had taken over from the Auto Architect, Gerald's influence may have become less evident to the car buying public. But, in an industry where engineers and managers were acquainted they drew on each other's knowledge and experience, as for instance when Rover sought information on the small Buick V8 (see page 116). Gerald's overall contribution was extensive and younger former colleagues recently contacted remember him always on time at Luton, even though he had the longest distance to travel from Iffley, usually with a ready smile and with concern for the welfare of others.

Index

Illustrations and references in captions shown in italics

AC Cars 27, 57
Alcan Group 45
Alden, John *110,* 121, 123
Alfa Romeo 79, 98, 125
Alvis 57, 99, 102, 141, 143, 144
Armstong Siddeley 49
Aston Martin 27
Audi 144
Austin 7, 23, 37, 52, 67, 69, 74, 96, 99, 101, 103, 138
Austin, Sir Herbert 67

Baldwin, John 53
Barson, Chalenor 25, 29, 30
Bartlett, Charles 50
Bedford 118, *121,* 123, 130
Bentley 71, 125
Bertone 79
Birla Brothers 93
Black, Sir John 52
BMC ohc engine in EX181 *75,* 141
BMW 30, 143-145
Boden, Oliver 67
Breen, Terry 130
Briggs Motor Bodies 39, 50-52
Bristol 57
British Leyland 103, 145
British Motor Corporation (BMC) 36, 67, 69, 86, 99, 102, 105, 117, 137, 138, 141, 143-145
British Power & Light 49
Bugatti 71, 125, *127,* 130, *131,* 135
Buick 113, 115, 116
Burzi, Dick 101, 143
Buxted Chicken Company 29

Castro, Fidel 113
Chappel Piano Co. 36
Chase, Pam. 51
Chayne, Charles 112
Chevrolet 106, 113, 115
Chrysler 46
Churchill, Winston 33
Ciba Geigy 36
Citroen 36, 52, 109

Clark, Henry Austin 113, 130
Clark, Jimmy 113
Clausager, Anders 137
Cleveland, F.J. 25
Clore, Charles 49
Cole, Ed. 114
Cooke, Tony 117
Copelin, Philip 109, 119
Cousins, Cecil 141, 142
Crowthorne Road Research Labs. 124
Curtice, Harlow (Red) *111,* 112
Cyclemaster 51

Daimler, Gottlieb 109
Daimler Motors 81, 129
Daimler Paul 131
Daniels, Jack 102, 143
Darder, Andres 121
De Havilland 25, 31
Delage 125
De La Rue 34
Delorean, John Z. 115
Department of Transport 123
Deroy 25-30, *26-30,* 31, 47, *47, 48,* 98, 142, 145
Dono, George 45, 85
Downs, Dickie 124
Drew, Harold 109, 115
Dunlop Co. 86

Earl, Harley 111
Eberhorst, Eberan von 64, 66
Eminson, Dr. Clarence *139*
Enever, Syd. 78, 141
Eustace Watkins 107

Farina, Pinin 71
Fedden Prototype 42, *42*
Fedden, Sir Roy 39, 52
Ferguson, Harry 101
Fiat 102, 144
Fisher, Anthony G.A. 25, 29
Fisher Body Division 117
Flynn, Erroll 56, 59
Ford Lotus Cortina 77

147

Auto-Architect

Ford Model T 11, *11, 12*
Ford Motor Co. 102, 113, 118, 145
Frère, Paul 64

Gabo, Naum 38, 39
Gatsonides, Maurice 63
General Motors 23, 29, 52, 68, 109-111, *111*, 114-118, 120, *120*, 145
General Motors Technical Centre 113, 116
Girling 86, 88
Goldsmith, Max 102, 109
Gomm, Joe 73, 74
Gordon England, Eric C. 37
Grandfield, Charles 52, 93
Graves, Jo 106
Griffiths, Paul 98
Grimley, Horace 46, 63

Hall, Leslie 31
Hanks, Reginald F. 67, 99, 102, 103, 139, 140
Harriman, Sir George 101, 103
Harrison, T.C. (Cuth) 61, 62, 64, *65*
Harvey, Albert 98
Hayes gearbox 23
Hennessy, Patrick 102, 109
Hepworth & Grandage 130
Hill, Phil 77
Hobbs, Wilfred 102, 103
Hodgson 98
Holmes, Fred 106
Hotchkiss 138
Hurlock, Charles 27

Illingworth 46
Indianapolis 500-mile Race 113, 29
Institute of Auto. Eng. (I.A.E.) 13, 23, 27, 29, 31, 109, 131
Issigonis, Alec 33, 36, 67, 68, 99, 102, 106, 109, 139, 141, *142*, 143-145

Jackman, Bernard 116
Jackson, Brian 87
Jaguar 81
J.A.P. 16
Jeddere-Fisher, Arthur 112, 130
Johnson, Eric 135
Jones, David *110*, 112, 117, 119, 145
Jowett, Benjamin 37
Jowett Bradford van 37, 39, 51, 52, 53, *53*, 57, *57*
Jowett Car Club 124
Jowett Company 36, 37, 42, 45, 51, 53, *56-58*, 64, 66, 93, 94, 95, 98, 113, 121
Jowett Javelin 37-46, *40-44, 46*, 49, 50, 52, 53, *53-55*, 55, *56-60*, 57, *62, 63*, 63-66, *65*, 101, 105, 113, *123*, 130, 137, 138, *139*, 145
Jowett Jupiter 64, *66*
Jowett, William 37, 46

Kensington-Moir, Bertie 101
Kimber, Cecil 31
Kinchin, Joe 86
Knowles, David 141
Knox 13
Knudson, 'Bunky' 97
Korner, Reg. 42

Lanchester, Frederick W. 23, 27, 105
Lancia 56, 106
Land Rover 96, 116
Larkin, Arthur 122
Laystall 122
Lazard, (Bankers) 49, 66
Le Mans 1951 *66*
Libby Owens Ford Company 114
Lincoln 39
Lockheed 89, 116
Lord, Leonard 50, 67, 99, 101, 102, 106, 137, 143, 144
Lucas 123
Lucas, Ralph 16, 17
Lunn, Roy 113
Lyons, Sir William 109

Macintosh, Robert 34, 36
Maher, Eddie 77,
Majorca (cottage) 121, *122*
Marryat & Scott 24
Martin, David *74*
Mays, Raymond 112, *132*
McKay, Ken 87
Meadows 66
Mercedes-Benz 81, 112, 113, 121, 124, 125, 129, 130, *130*, 131, *131-134*, 133, 135, 136, 145
Metalastik 102
Metcalf 46
M.G. Car Club 124
M.G. Car Company 67, 69, 125
M.G. EX181 *75*, 77

M.G. Magnette Mark lll 143
M.G. MGA 75, 77, 80, *89*, 141
M.G. Prototypes *77-79*, 141
M.G. TD/TF 74, 80, 138
M.G. Y-type 31
M.G. Z Magnette 71, *72*, *73*, 74, 76, 77, 81, *89-91*, 101, 109, 117, 120, *123*, 137-141, 145
Mini-Minor 68, 106, *142*
Mini-Minor 9X replacement 144, *144*
Monaco GP *134*, 135, 136
Monte Carlo Rally 61, 64, *65*
Montgomery, Field Marshal 52
Morphy Richards 49
Morris Bodies Branch 78
Morris Drawing Office 31
Morris Engines Branch 77
Morris, John 36
Morris Motors 25, 31, 32, 38, 67-69, 99-104, 138, 139
Morris, William *see Nuffield, Lord*
Moss, Stirling *76*, 77, 136
Moulton, Alex 102, 143, 144

Nader, Ralph 115
NADEX exhibition 128
Napier 22
North Lucas Radial 14, *14*, *15*, 16, 17, 39
North, Oliver D. 11, 16, 17, 20, 22, 23, 39, 143
Nuffield Group 31, 49, 57, 66, 85, 88, 93, 101, 103, 106, 138
Nuffield, Lord (William Morris) 31, 34, 67, 102, 137

Oak, A.Vic. 31, 34, 67, 68, 77, 85, 99, 139, 141, 144
Oldsmobile 114, 116
O'Neill, Jim 138, 139, *140*
Opel 117, 118, 120, 122, 123, 145
Opel Kadett 119, 120
Orchard House 93, 94, *97*, 98, 110
Oxford Hoist *126*, 127, 128
Oxford Vaporiser 31, *32*, *33*, 36

Palmer, Celia 38, 56
Palmer, Diana 24, *24*, 38, 98, 102, *111*, 129, 131, 134, *134*, 135, 145
Panhard 27
Patthey, Hubert 57
Pauley, Reg. 122

Payne, F. J. Ltd. 125, 126
Payne, John 125, 127, 128, 129, 130, 131
Perkins 92, 95, 96
Perkins, D. *110*
Petter *94*, 96,
Peugeot 102, 105, 144
Platt, Maurice 109, *110*, 112-114, 117, 119, 121
Pomeroy, Laurence 55, 56, 130
Pontiac *112*, 115
Poole, Steven 39, 66
Porsche, Ferdinand 39, 129
Prescott Hill Climb *133*
Pressed Steel Co. 73
Pressnell, Jon 140
Prix Maritime *133*

Raffay, Gerhardt von *133*
Read, Herbert 39
Reckitt & Coleman 29
Reilly, Charles Calcott (Peter) 36-39, 49, *50*, 51, 55, 66
Renault 102
Renfrew Foundries 45
Reyrolle 49
Ricardo 124
Richardson, Joy 108
Richardson, Peter 106, 108
Richmond, Joan 25
Riddle, Edward 16
Ridley, Lord 112
Riley, Ed. 109, 112, 113
Riley Motors 55, 67, 77, 88, 105, 139, 142, 143, 145
Riley Pathfinder 81, *82-85*, 87-89, *89*, 100, 141, 143-145
Rix, John 99
Robbins, Brian 31
Rolls-Royce 45, 110
Rolt, Tony 101
Rootes, Reginald 50
Rover 116, 146
Rowlands, David 87, 141
Rush, Harry 67, 68, 139
Russell, Dr. Ritchie 127

Saab 27
Sainsbury, Wilfred 49
Salter, Frank 52
Sampietro, Achille 93

Auto-Architect

Saxon 10
Scammell 13, 16, *17-22*, 20, 22, 24, 25, 27, 47, 48
Schuster, Gerd 145
Scripps Booth 11, *11*
Shelsley Walsh *132*
Shuttleworth Collection 130
Silverstone 131, *133*
Smith, Ron 63
Smith, S.V. 142
S.M.M.T. 53, 123
Spa 24-hour race 64, *65*
Standard 52
Steyr 39
Stierli, Joseph 57
Stokes, Donald 145
Stott, Tom *110*, 117
S.U. Carburettor Co. 36
Sutherland, R. Gordon 27
Sutton, Jill 135
Sutton, John 135

Talbot 25
Targa Florio race 112, 129, *133*
Taylor, Ken 94
Thomas, Miles 50, 67, 93, 103
Thompson, John Ltd. 85, 88, 142
Thomson, Jimmy 77
Thomson & Taylor Ltd. 93, 94, 96
Thornley, John *74*, 140, 141, 142
Tjaarda, John 39
Tothill, Peter 73, 88
Triplex 52

V4 engine for B.M.C. *104*, 106
Vandervell, Tony 130
Vauxhall Motors 14, 29, 66, 68, 85, 97, 109, *110*, 112, 114, 115, 117-120, *120*, 121, 123-125, 130, 135, 141, 143, 145
Vauxhall Velox/Cresta 111, 122
Vauxhall Ventora 122
Vauxhall Victor 110, 113, 114, 117, 118, *118-119*, 119, 120, 122, 145
Vauxhall Viva 110, 117, *118*, 119, 120
Vauxhall VX 4/90 *118*, 122
Volkswagen 39, 42, 44, 93, 114, 115
Volvo 110

Wansbrough, George 49, 51, *54*, 55, 93, *95*, 97
Wansbrough, Nancy 121
Webster, Harry 145
Werner, Christian 129, 135
White Motor Corporation 97
Wilks, Spencer 50
Wilson, Charles 112
Wilson, Harold 50
Wisdom, Tommy 64, *65*
Wise, Tommy 61-64, *65*
Wolseley Motors 67, 69, 70, 105, 138, 139
Wolseley 4/44 *68-70*, 70, 73, 74, 89, 117, *123*, 138, 140, 141
Wolseley 6/80 100
Wolseley 6/90 81, 89, 100, *100*, 101, 106-108, *108*, *123*, 138, 141-144
Woodhead, Harry 49, 50

Yeoman *92*, 93-97, *94-96*

150